In Memoriam
Thomas Lehmkuhl
1957–2004

MEMOIRS
of the
American Mathematical Society

Number 921

# Compactification of the Drinfeld Modular Surfaces

Thomas Lehmkuhl

January 2009 • Volume 197 • Number 921 (third of 5 numbers) • ISSN 0065-9266

**American Mathematical Society**
Providence, Rhode Island

2000 *Mathematics Subject Classification.*
Primary 11G09, 13D10, 14B20.

**Library of Congress Cataloging-in-Publication Data**

Lehmkuhl, Thomas.
Compactification of the Drinfeld modular surfaces / Thomas Lehmkuhl.
p. cm. — (Memoirs of the American Mathematical Society, ISSN 0065-9266 ; no. 921)
"Volume 197, number 921 (third of 5 numbers)."
Includes bibliographical references and index.
ISBN 978-0-8218-4244-7 (alk. paper)
1. Drinfeld modules. 2. Deformations (Mechanics) 3. Surfaces, Algebraic. I. Title.

QA247.3.L44 2009
512′.42—dc22 2008039489

## Memoirs of the American Mathematical Society

This journal is devoted entirely to research in pure and applied mathematics.

**Subscription information.** The 2009 subscription begins with volume 197 and consists of six mailings, each containing one or more numbers. Subscription prices for 2009 are US$709 list, US$567 institutional member. A late charge of 10% of the subscription price will be imposed on orders received from nonmembers after January 1 of the subscription year. Subscribers outside the United States and India must pay a postage surcharge of US$65; subscribers in India must pay a postage surcharge of US$95. Expedited delivery to destinations in North America US$57; elsewhere US$160. Each number may be ordered separately; *please specify number* when ordering an individual number. For prices and titles of recently released numbers, see the New Publications sections of the *Notices of the American Mathematical Society*.

**Back number information.** For back issues see the *AMS Catalog of Publications*.

Subscriptions and orders should be addressed to the American Mathematical Society, P. O. Box 845904, Boston, MA 02284-5904, USA. *All orders must be accompanied by payment.* Other correspondence should be addressed to 201 Charles Street, Providence, RI 02904-2294, USA.



*Memoirs of the American Mathematical Society* (ISSN 0065-9266) is published bimonthly (each volume consisting usually of more than one number) by the American Mathematical Society at 201 Charles Street, Providence, RI 02904-2294, USA. Periodicals postage paid at Providence, RI. Postmaster: Send address changes to Memoirs, American Mathematical Society, 201 Charles Street, Providence, RI 02904-2294, USA.


This publication is indexed in *Science Citation Index*®, *SciSearch*®, *Research Alert*®,
*CompuMath Citation Index*®, *Current Contents*®/*Physical, Chemical & Earth Sciences.*
Printed in the United States of America.

∞ The paper used in this book is acid-free and falls within the guidelines
established to ensure permanence and durability.
Visit the AMS home page at `http://www.ams.org/`

10 9 8 7 6 5 4 3 2 1 14 13 12 11 10 09

# Contents

Introduction ix

Chapter 1. Line Bundles 1
1. Basic notions 1
2. Homomorphisms of line bundles 3
3. Quotients 7
4. The convergence lemma 10

Chapter 2. Drinfeld Modules 15
1. Analytical definition of Drinfeld modules 15
2. The category of Drinfeld modules 19
3. Drinfeld modules over fields 22
4. Level structures 25
5. Modular manifolds 28
6. Pseudo-Drinfeld modules 30

Chapter 3. Deformation Theory 33
1. Deformations of Drinfeld modules 33
2. Deformations of isogenies 40
3. Deformations of level structures 42
4. Smoothness of the moduli spaces 46
5. Group action on the moduli space 47

Chapter 4. Tate Uniformization 51
1. Formal schemes 51
2. Good and stable reduction 54
3. Lattices and Tate data 57
4. Group action 64

Chapter 5. Compactification of the Modular Surfaces 67
1. Formal representability of Tate data 67
2. The universal Drinfeld module with bad reduction 73
3. Algebraization 78

Appendix 85
A. Induced schemes 85

B. Construction of coherent sheaves 86
Bibliography 89
Glossary of Notations 91
Index 93

… Abstract

In this article we describe in detail a compactification of the moduli schemes representing Drinfeld modules of rank 2 endowed with some level structure. The boundary is a union of copies of moduli schemes for Drinfeld modules of rank 1, and its points are interpreted as Tate data. We also study infinitesimal deformations of Drinfeld modules with level structure.

Received by the editor March 22, 2004; and in revised form July 25, 2006.

2000 *Mathematics Subject Classification.* Primary 11G09, 13D10, 14B20.

*Key words and phrases.* Drinfeld module, level structure, deformation, lattice, formal schemes.

# Introduction

The introduction of elliptic modules by V.G. Drinfeld in the seventies, [**Dr**], now called Drinfeld modules, was the starting point of a widely ramified theory concerning function fields over finite fields. Within leh-abstract the analogy between function fields and number fields, Drinfeld modules are the analogues to elliptic curves. Many aspects of elliptic curves turned out to have a correspondent in the theory of Drinfeld modules, e.g. transcendence, diophantic approximation, modular forms, etc.. Good references for these various aspects are the collections [**GPRV**], [**GHR**], [**KLS**], and the monographs [**Go**] and [**La**], of course. A further reference is [**Ge1**].

Let $A$ be the coordinate ring of a fixed smooth projective curve minus one closed point. Then Drinfeld modules (of a rank $d$) are defined over base schemes $S$ lying over $\operatorname{Spec} A$. They give rise to moduli problems which are representable by a smooth algebraic stack. In order to get representing schemes it is necessary to impose level $I$ structures on them for a suitable ideal $I$ of $A$. Let $M_I^d \to \operatorname{Spec} A$ be the corresponding moduli scheme. According to Drinfeld, this is a regular scheme of dimension $d$. It is subtle to show this for points lying over $V(I)$ (case of characteristic within the level). The original way is to study deformations of (non-local) formal groups with level structure. One shows that these are represented by a regular local ring. After that one constructs a map from deformations of Drinfeld modules to those of formal groups and proves that this is a bijection. In [**Dr**], §9, Drinfeld constructs a relative compactification of the modular surfaces $M_I^2$. But the presentation is rather sketchy, and a detailed exposition is still missing.

The aim of this work is to fill this gap. Briefly, the work is organized as follows.

In chapter 1, we discuss line bundles considered as algebraic groups and their endomorphisms. Roughly speaking, these are polynomials of the Frobenius endomorphism. After that we study commuting endomorphisms. It turns out that their coefficients have to satisfy strong conditions. Similarly, we consider the case of additive formal power series. The main result is the convergence lemma proved in section 4.

In chapter 2, we introduce the concepts of Drinfeld modules, of isogenies, and of level structures emphasizing the case of characteristic within the level. If the level $I$ contains at least two points, Drinfeld modules of level $I$ lead to a representable module problem. Its moduli space is constructed in section 5. So far, all results are taken from [**Dr**] supplied with some details of proofs. These two chapters are a self contained introduction to Drinfeld modules. We have included them here in spite of the references cited above, firstly for the reader's convenience, and secondly, because the case of characteristic within the level is not treated in loc. cit..

Chapter 3 is devoted to the infinitesimal deformation theory of Drinfeld modules. We follow the cohomological approach, given by [**La**] for the case of Drinfeld modules of characteristic away from the level. In order to extend it to the general case, we consider deformations of isogenies. After that we are able to show Drinfeld's theorem on the smoothness of the moduli spaces avoiding deformations of formal groups. As an application of deformation theory, we include a proof of Gekeler's de Rham isomorphism theorem, [**Ge**]. In the last section, we describe the action of the adelic group $\mathrm{GL}_d(\mathbb{A}_\mathrm{f})$ on $M^d := \varprojlim_I M_I^d$, and the important fact that $M_I^d$ is the quotient of $M^d$ by the congruence subgroup $\Gamma(I) := \ker(\mathrm{GL}_d(\hat{A}) \to \mathrm{GL}_d(A/I))$.

Next we turn to the boundary of $M_I^d$. The method is to consider morphisms from the generic point $\operatorname{Spec} K$ of a discrete valuation ring $O$ which cannot be extended to the whole scheme $\operatorname{Spec} O$. As an intermediate step to compactification, we introduce the concept of a formal boundary. This is a formal scheme, which can be thought of as the completion of a compactification yet to be constructed, along the boundary. In the present case of $M_I^d$, these morphism have a modular interpretation as *Tate data* with level structure. This suggests that the boundary should have a stratification with strata isomorphic to $M_I^{d_1}$ for $d_1 < d$. This was confirmed by Kapranov, [**Ka**], in some special cases.

In the last chapter we construct in case $d = 2$ a formal boundary of $M_I^2$, whose special fiber is a disjoint union of spaces isomorphic to $M_I^1$. Then we want to show that this formal scheme derives from an actual scheme. For this purpose, we prove a general glueing result. To apply this in our situation, we construct an analog to the Tate curve in the theory of elliptic curves. Using this, we succeed in establishing the crucial isomorphism needed for glueing.

As mentioned above, the material of this work is not new. Nearly all is contained in Drinfeld's fundamental original paper. The author's contribution is to explain some details of the beautiful ideas contained in it. During the final preparation of this paper G.-J. van der Heiden published another treatment of Drinfeld's compactification, which puts emphasis on the Weil pairing and is different from the approach presented here.

This monograph is the revised version of the author's Habilitationsschrift [**Le**]. It was begun while the author was guest at the Sonderforschungsbereich 170 at the university of Göttingen. I am grateful to the Mathematische Institut Göttingen for support and hospitality. I am particular indebted to U. Stuhler for constant encouragement and interest.

Thomas Lehmkuhl

*We were deeply saddened when the author, Thomas Lehmkuhl, died in November 2004 after many years of illness, which he had taken in great spirit.*

*The present paper is his "Habilitationsschrift". It was submitted by him for publication in the Memoirs of the AMS on the 22th of March, 2004.*

*For refereeing the paper and for a very carefull reading we thank Yakov Varshavsky from the Hebrew University of Jerusalem.*

*We did a final reading of the paper in June/July 2006, so all remaining points are our responsibility.*

*We would like to thank also Dan Abramovich for support as an editor of the Memoirs in this sad and difficult situation.*

Ulrich Stuhler Stefan Wiedmann

CHAPTER 1

# Line Bundles

Drinfeld modules are structures on line bundles over a base scheme of positive characteristic. The latter are commutative algebraic groups by means of the usual addition. In this chapter we collect some of their basic properties and study homomorphisms between them. The basic example is the Frobenius endomorphism. All other endomorphisms of a line bundle are linear combinations of powers of the Frobenius. Moreover, we study commuting endomorphisms. It turns out that this imposes strong conditions on their "coefficients". References are [**Go, La, GPRV**].

## 1. Basic notions

We refer to [**Ha**], chap. 2, for the basics of algebraic geometry, in particular for basic facts on schemes.

In this paper, all schemes have characteristic $p$, where $p$ is a fixed prime number. Let $f : X \to Y$ be a morphism of schemes, then

$$f^\sharp : \mathcal{O}_Y \to f_*\mathcal{O}_X$$

denotes the morphism of their structure sheaves. The same symbol is used for a datum that determines such a morphism, e.g. the section of $\mathcal{O}_X(X)$ defining a morphism $f$ to $Y = \operatorname{Spec} \mathbb{F}_p[T]$ is named $f^\sharp$.

$\mathbb{A}^1_S$ is the affine line over the scheme $S$.

For an affine scheme we frequently write $B$ instead of $\operatorname{Spec} B$.

The cardinality of a set $M$ is denoted by $|M|$.

Let $\mathcal{L}$ be an invertible sheaf on $S$. The *line bundle* corresponding to $\mathcal{L}$ is the $S$-scheme

$$\mathbb{G}_{a,\mathcal{L}} := \operatorname{Spec}\ \mathrm{S}_{\mathcal{O}_S}(\mathcal{L}^{-1})\,.$$

Here S denotes the symmetric algebra, c.f. [**Ha**], II, ex. 5.16 and 5.17. It is a commutative group scheme over $S$. We denote by *add* its addition. Then $add^\sharp$ is the homomorphism defined by the diagonal map

$$\mathcal{L}^{-1} \to \mathcal{L}^{-1} \oplus \mathcal{L}^{-1} \subseteq \mathcal{O}_{\mathbb{G}_{a,\mathcal{L}} \times \mathbb{G}_{a,\mathcal{L}}}\,.$$

There is a functorial isomorphism of groups

$$\mathbb{G}_{a,\mathcal{L}}(T) \cong \Gamma(T, \mathcal{L} \otimes_{\mathcal{O}_S} \mathcal{O}_T)^+$$

for any $S$-scheme $T$. If $S = \operatorname{Spec} R$ and $\mathcal{L} \cong \mathcal{O}_S$, then a trivialization provides an isomorphism

$$\mathbb{G}_{a,R} \cong \operatorname{Spec} R[X] ,$$

and the addition is given by $add^\sharp(X) = X + Y \in R[X, Y]$. We will always consider $\operatorname{Spec} R[X]$ as a group in this way.

Let $S$ be a scheme (over $\operatorname{Spec} \mathbb{F}_p$). Then there is the *absolute Frobenius endomorphism*

$$\operatorname{Frob}_S : S \to S ,$$

which is the identity on topological spaces and has $\operatorname{Frob}_S^\sharp(t) = t^p$ for any section $t$.

Let $X \to S$ be a scheme over $S$. Then there is a factorization of the Frobenius of $X$:

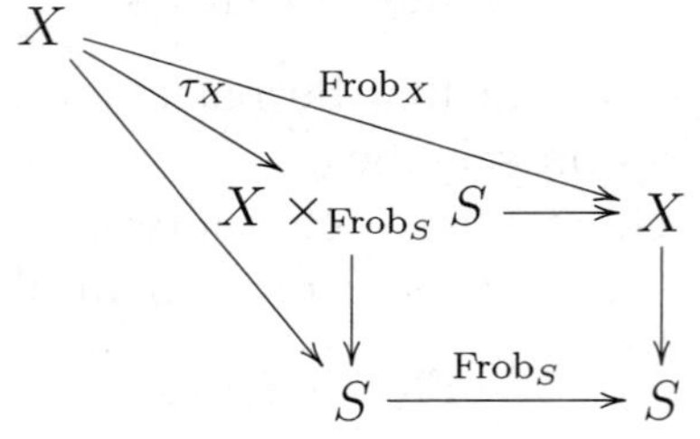


Here the square is cartesian. The morphism $\tau_X$ is called the *relative Frobenius morphism* of $X$.

EXAMPLES 1.1. (1) Let $S = \operatorname{Spec} R$ and let $X = \operatorname{Spec} R[T]$. Then $X \times_{\operatorname{Frob}_S} S$ is canonically isomorphic to $X$, and $\tau_X$ is the morphism given by $\tau_X^\sharp(T) = T^p$

(2) Let $\mathcal{L}$ be an invertible sheaf over $S$, and let $X = \mathbb{G}_{a,\mathcal{L}}$ be the line bundle of $\mathcal{L}$. Then $\tau_X^\sharp$ comes from the isomorphism $(\mathcal{L}^{-1})^\tau := \operatorname{Frob}_S^*(\mathcal{L}^{-1}) \cong (\mathcal{L}^{-1})^{\otimes p}$ given by $1 \otimes f \mapsto f \otimes \ldots \otimes f$.

If $q = p^m$, we denote by $\operatorname{Frob}_{S,q}$ the iterated Frobenius $\operatorname{Frob}_S^m$. Similarly there is a relative Frobenius $\tau_{X,q}$, which we denote by $\tau_q$. With the identification made in example (1), we have $\tau_q = \tau_p^m$ for $X = \operatorname{Spec} R[T]$. On the other hand, if $X = \mathbb{G}_{a,\mathcal{L}}$, where $\mathcal{L}$ is isomorphic to $\mathcal{O}_S$ , then an identification $X \times_{\operatorname{Frob}_R} \operatorname{Spec} R \cong X$ depends on the choice of a trivialization of $\mathcal{L}$.

## 2. Homomorphisms of line bundles

Let $S$ be a scheme, and let $\mathbb{G}_{a,\mathcal{L}}$ and $\mathbb{G}_{a,\mathcal{M}}$ be line bundles over $S$. A *homomorphism* from $\mathbb{G}_{a,\mathcal{L}}$ to $\mathbb{G}_{a,\mathcal{M}}$ is a homomorphism of group schemes. The set of homomorphisms from $\mathbb{G}_{a,\mathcal{L}}$ to $\mathbb{G}_{a,\mathcal{M}}$ is an abelian group denoted by $\mathrm{Hom}(\mathbb{G}_{a,\mathcal{L}}, \mathbb{G}_{a,\mathcal{M}})$. If $\mathcal{L} = \mathcal{M}$, it is even a ring. Our aim is to compute these groups. First consider the local case $\mathbb{G}_{a,\mathcal{L}} = \mathbb{G}_{a,\mathcal{M}} = \mathbb{G}_{a,R} = \mathrm{Spec}\, R[X]$, where $R$ is a ring. In this case, the relative Frobenius $\tau_p$ is canonically an endomorphism, c.f. 1.1, (1). Multiplication by elements of $R$ gives additional endomorphisms. In fact, we can prove:

PROPOSITION 2.1. *The endomorphism ring of $\mathbb{G}_{a,R}$ consists of all polynomials*

$$\sum_{i\geq 0} a_i\tau^i,\ a_i \in R,$$

*where $\tau = \tau_p$. One has the commutation relation*

$$\tau \cdot a = a^p \cdot \tau \quad \text{for } a \in R.$$

PROOF. Let $\varphi$ be a morphism from the $R$-scheme $\mathrm{Spec}\, R[X]$ to itself. It is a homomorphism if and only if $\varphi^\sharp(X+Y) = \varphi^\sharp(X)+\varphi^\sharp(Y)$. Writing $\varphi^\sharp = \sum a_i X^i$, this is true if and only if $a_i(X+Y)^i = a_iX^i+a_iY^i$. Hence we must have $\gcd(\binom{i}{j}, 1 \leq j < i)\cdot a_i = 0$. By the next lemma, this is equivalent to the condition that $a_i$ vanishes for all $i$ that are not a power of $p$. This proves the first statement. The second one is obvious. ☐

LEMMA 2.2. *Let $n$ be an integer $> 1$. Then the gcd of $\binom{n}{k}$, $0 < k < n$, is 1 if $n$ is not a power of a prime and $\ell$ if $n = \ell^r$, where $\ell$ is a prime.*

PROOF. Let $\ell$ be a prime and $v_\ell(m)$ the (additive) valuation of $m$ at $\ell$. Then we have

$$v_\ell(m!) = \sum_{i=1}^{\infty} \left\lfloor \frac{m}{\ell^i} \right\rfloor.$$

Hence

$$v_\ell\left(\binom{n}{k}\right) = \sum_{i=1}^{\infty}\left(\left\lfloor\frac{n}{\ell^i}\right\rfloor - \left\lfloor\frac{k}{\ell^i}\right\rfloor - \left\lfloor\frac{n-k}{\ell^i}\right\rfloor\right).$$

Let $n = m \cdot \ell^r$ with $(m, \ell) = 1$. If $m > 1$, then we get for $k = \ell^r$:

$$v_\ell\left(\left(\begin{array}{c} m\ell^r \\ \ell^r \end{array}\right)\right) = \sum_{i=1}^{r} (m \cdot \ell^{r-i} - \ell^{r-i} - (m-1)\ell^{r-i}) + \sum_{i=r+1}^{\infty} \left(\left\lfloor \frac{m}{\ell^{i-r}} \right\rfloor - \left\lfloor \frac{m-1}{\ell^{i-r}} \right\rfloor\right) = 0.$$

Applying this for all prime numbers $\ell$, gives the statement of Lemma 2.2. in case, that $n$ is not a power of a prime.

On the other hand, if $n = \ell^r$, $v_\ell\left(\left(\begin{array}{c} \ell^r \\ k \end{array}\right)\right) \geq 1$ for $0 < k < \ell^r$; for $k = \ell^{r-1}$ we find

$$v_\ell\left(\left(\begin{array}{c} \ell^r \\ \ell^{r-1} \end{array}\right)\right) = \sum_{i=1}^{r} \ell^{r-i} - \sum_{i=1}^{r-1} \ell^{r-1-i} - \sum_{i=1}^{r-1} \ell^{r-1-i}(\ell - 1) = 1.$$

Furthermore, $\left(\begin{array}{c} \ell^r \\ k \end{array}\right)$ is divisible by $\ell$ for all $k$, $0 < k < n = \ell^r$. Finally, $\left(\begin{array}{c} \ell^r \\ 1 \end{array}\right) = \ell^r$ is not divisible by any other prime than $\ell$. The result follows from this. □

The polynomials $f \in R[X]$ satisfying $f(X+Y) = f(X) + f(Y)$ are called *additive*. The same definition applies to power series $f \in R[[X]]$.

The endomorphism ring of $\mathbb{G}_{a,R}$ is denoted by $R\{\tau_p\}$. This ring could be defined formally by adjoining an indeterminate $\tau$ to $R$ with commutation rule

$$\tau \cdot a = a^p \cdot \tau\,.$$

In the literature, this ring is called the *skew-polynomial ring.*

Let $q = p^m$, and let $R$ be a $\mathbb{F}_q$-algebra. Then an endomorphism $\varphi \in R\{\tau_p\}$ is called $\mathbb{F}_q$-*linear*, if $a\varphi = \varphi a$ for all $a \in \mathbb{F}_q$. It is easy to see that the ring of $\mathbb{F}_q$-linear endomorphisms is just $R\{\tau_q\} \subseteq R\{\tau_p\}$, which is defined in the obvious way.

Although $R\{\tau\}$ is not commutative in general, it shares many properties with the ordinary polynomial ring. For example, if $R$ is a perfect field, it admits a (right and left) euclidean algorithm, see [**Go**] for details.

Now we turn to the global case. Let $S$ be a scheme, and let $\mathcal{L}$, $\mathcal{M}$ be invertible sheaves on $S$. From the fact that morphisms $\mathbb{G}_{a,\mathcal{L}} \to \mathbb{G}_{a,\mathcal{M}}$ (considered as schemes over $S$) correspond bijectively to homomorphisms $\mathcal{M}^{-1} \to \bigoplus_{i\geq 0} \mathcal{L}^{-i}$ of locally free sheaves or equally to sections in $\Gamma(S, \bigoplus_{i\geq 0} \mathcal{M} \otimes \mathcal{L}^{-i})$, and from our discussion of the local case, one deduces immediately

PROPOSITION 2.3. *The homomorphisms from $\mathbb{G}_{a,\mathcal{L}}$ to $\mathbb{G}_{a,\mathcal{M}}$ correspond bijectively to the set of sections of the sheaf $\bigoplus\limits_{n\geq 0} \mathcal{M} \otimes_{\mathcal{O}_S} \mathcal{L}^{-p^n}$.*

If $\varphi$ is a homomorphism of line bundles, we denote by $\tilde{\varphi} = \sum_n \tilde{\varphi}_n$ the corresponding section.

REMARK 2.4. Using the canonical isomorphism

$$\mathcal{L}^{-p^n} \cong (\mathcal{L}^{-1})^{\tau^n} = \mathcal{L}^{-1} \otimes_{\mathrm{Frob}_S^n} \mathcal{O}_S\,,$$

and writing $\tau_p^n$ for the $n$-th iterated relative Frobenius, any homomorphism $\varphi$ can be written as

$$\varphi = \sum_{n\geq 0} a_n \tau_p^n\,,$$

where $a_n$ is the homomorphism from $\mathbb{G}_{a,\mathcal{L}^{\tau^n}}$ to $\mathbb{G}_{a,\mathcal{M}}$ that corresponds to $\tilde{\varphi}_n$, and the sum is locally finite.

Let $S$ be a scheme over $\mathbb{F}_q$, $q = p^m$. By $\mathrm{End}_{\mathbb{F}_q}(\mathbb{G}_{a,\mathcal{L}})$ we denote the sub-algebra of $\mathbb{F}_q$-linear endomorphisms. It consists exactly of those endomorphisms $\varphi$, whose section $\tilde{\varphi}$ have only components in $\Gamma(S,\, \mathcal{L}^{1-q^n})$, $n \geq 0$. Writing $\tau_q^n = \tau_p^{mn}$ (see the remark above), they are the endomorphisms of the form $\sum_n a_n \tau_q^n$.

There is a canonical homomorphism

$$\partial : \mathrm{End}(\mathbb{G}_{a,\mathcal{L}}) \to \mathcal{O}_S(S)$$

sending $\varphi$ to $\tilde{\varphi}_0$ (or to $a_0$ with the notation of the remark). For $\varphi \in \mathrm{End}(\mathrm{Spec}\, R[X])$ we have $\partial\varphi = D\varphi^\sharp(0)$, where $D$ is the derivative.

In the following chapters we are in particular interested in those homomorphisms that are finite as morphisms of schemes. They are characterized by the following propositions.

PROPOSITION 2.5. *Let $\varphi : \mathbb{G}_{a,\mathcal{L}} \to \mathbb{G}_{a,\mathcal{M}}$ be a homomorphism. Then the following conditions are equivalent:*

(a) *$\varphi$ is flat;*
(b) *$\varphi$ is quasi-finite;*
(c) *the corresponding section $\tilde{\varphi} \in \Gamma(S, \bigoplus\limits_{i\geq 0} \mathcal{M} \otimes \mathcal{L}^{-p^i})$ is vanishing nowhere.*

PROOF. The question is local, so we may assume $S = \mathrm{Spec}\, R$ and $\mathcal{L}$ and $\mathcal{M}$ are trivial. Thus we have $\mathbb{G}_{a,\mathcal{L}} = \mathrm{Spec}\, R[X]$ and $\mathbb{G}_{a,\mathcal{M}} =$

$\operatorname{Spec} R[Y]$. The complex

$$0 \to R[Y][X] \xrightarrow{d} R[Y][X] \to 0\,,$$

where $d$ is the multiplication by $Y - \varphi^\sharp(Y)$, is a free resolution of $R[X]$ over $R[Y]$. Hence, using [Bou], chap II, §3, no. 4, Corollary after Prop. 15, $\varphi$ is flat if and only if for all maximal ideals $\mathfrak{m} \subseteq R[Y]$, $d \otimes_{R[Y]} R[Y]/\mathfrak{m}$ is injective. This is true if and only if the coefficients of $\varphi^\sharp(Y)$ generate the unit ideal. Hence (a) $\Leftrightarrow$ (c).

(b) $\Leftrightarrow$ (c) is obvious. □

PROPOSITION 2.6. *Let $\varphi : \mathbb{G}_{a,\mathcal{L}} \to \mathbb{G}_{a,\mathcal{M}}$ be a homomorphism and let $S$ be connected. Then the following conditions are equivalent:*

(a) *$\varphi$ is finite;*
(b) *if $\tilde{\varphi} = \sum_{i=0}^{m} \tilde{\varphi}_i$ is the corresponding section, then there exists a number $n \leq m$, such that $\tilde{\varphi}_n$ is an isomorphism and $\tilde{\varphi}_i$ is nilpotent for $i > n$, i.e. $\tilde{\varphi}_i \in \Gamma(S, \mathfrak{n}_S \otimes \mathcal{M} \otimes \mathcal{L}^{-p^i})$, where $\mathfrak{n}_S$ is the nilradical of $\mathcal{O}_S$.*

PROOF. Clearly (b) implies (a). If $\varphi$ is finite, it is flat by Prop. 2.5. Hence it has a rank, say $p^n$, because $S$ is connected. Therefore, $\tilde{\varphi}_n$ has no zero, and for $i > n$, $\tilde{\varphi}_i$ vanishes at each point of $S$. □

DEFINITION.

(a) If $\varphi$ is finite, the rank of the locally free $\mathcal{O}_{\mathbb{G}_{a,\mathcal{M}}}$-sheaf $\varphi_* \mathcal{O}_{\mathbb{G}_{a,\mathcal{L}}}$ is called the *degree* of $\varphi$, denoted by $\deg \varphi$. It is constant on connected components of $S$
(b) A homomorphism $\varphi$ is called *standard*, if it is finite and all $\tilde{\varphi}_i$, $i > n$ vanish.

PROPOSITION 2.7. *Let $\varphi$ and $\psi$ be standard endomorphisms of $\mathbb{G}_{a,\mathcal{L}}$ and $\mathbb{G}_{a,\mathcal{M}}$, and let $h$ be a homomorphism from $\mathbb{G}_{a,\mathcal{L}}$ to $\mathbb{G}_{a,\mathcal{M}}$ such that*

$$h \circ \varphi = \psi \circ h \quad .$$

*If $\deg \varphi > 1$ and $h \neq 0$, then $\deg \psi = \deg \varphi$ and $h$ is standard.*

PROOF. It is enough to check this locally in $S$. So we may assume $\mathbb{G}_{a,\mathcal{L}} = \operatorname{Spec} R[X]$ and $\mathbb{G}_{a,\mathcal{M}} = \operatorname{Spec} R[Y]$, where $R$ is a local ring.

Assume that not all coefficients of $h^\sharp = h^\sharp(Y)$ are nilpotent. Then there exists a prime ideal $\mathfrak{p} \in \operatorname{Spec} R$, such that $h^\sharp \not\equiv 0 \bmod \mathfrak{p}$ .

Since $\varphi$ and $\psi$ are standard, the equation $h \circ \varphi = \psi \circ h \bmod \mathfrak{p}$ implies $\deg \varphi = \deg \psi$. Let $p^d := \deg \varphi$, let $a, b$ be the leading coefficients of $\varphi^\sharp, \psi^\sharp$, respectively, and let $cX^{p^e}$ be the highest non vanishing term of

$h^\sharp$. Then, comparing leading coefficients of $(h \circ \varphi)^\sharp$ and $(\psi \circ h)^\sharp$, we obtain

$$c \cdot \left( c^{p^d - 1} - \frac{a^{p^e}}{b} \right) = 0 \,.$$

As $\varphi, \psi$ are standard, $a^{p^e}/b$ is a unit. We conclude that $c$ must be a unit, too.

Assume now that all coefficients of $h^\sharp$ are nilpotent. Then we may assume that $(h^\sharp)^p = 0$ but $h \neq 0$. Since $\varphi$ is standard, $(h \circ \varphi)^\sharp$ would have degree greater than $\deg(h^\sharp)$ whereas $(\psi \circ h)^\sharp$ has degree less or equal to $\deg(h^\sharp)$. This is a contradiction. □

PROPOSITION 2.8. *Let $\varphi$ be a finite endomorphism of $\mathbb{G}_{a,\mathcal{L}}$. Then there exists a unique automorphism $u$ of $\mathbb{G}_{a,\mathcal{L}}$, such that $\partial u = 1$ and $u \circ \varphi \circ u^{-1}$ is standard.*

PROOF. The question is local in $S$, so we may suppose $\mathbb{G}_{a,\mathcal{L}} = \mathbb{G}_{a,R}$. Moreover, we may assume that $\varphi$ mod $\mathfrak{n}'$ is standard for some ideal $\mathfrak{n}'$ satisfying $\mathfrak{n}'^2 = 0$.

Let $\varphi = \sum_{i=0}^{n} a_i \tau_p^i$ with $p^n > p^d := \deg \varphi$. Then

$$\varphi' := \left( id - \frac{a_n}{a_d^{p^{n-d}}} \tau_p^{n-d} \right) \circ \varphi \circ \left( id - \frac{a_n}{a_d^{p^{n-d}}} \tau_p^{n-d} \right)^{-1}$$

has leading term $b\tau_p^m$, where $m < n$. Using induction this proves the existence of $u$.

If $v$ is another automorphism as specified above, we put $h = v \circ u^{-1}$. Then we have

$$h \circ (u \circ \varphi \circ u^{-1}) = (v \circ \varphi \circ v^{-1}) \circ h \,,$$

so $h$ is standard by the last proposition. Consequently, we have $h = id$. □

## 3. Quotients

Let $H$ be a group scheme over a scheme $S$ acting on an $S$-scheme $X$ by

$$\omega : H \times_S X \to X \,.$$

Recall (c.f. [MFK], 0.6) that a *geometric quotient* is a morphism of $S$-schemes $\varphi : X \to Y$ satisfying the following conditions:

(i) The diagram

$$\begin{array}{ccc} H \times_S X & \xrightarrow{\;\omega\;} & X \\ \downarrow{\scriptstyle p_2} & & \downarrow{\scriptstyle \varphi} \\ X & \xrightarrow{\;\varphi\;} & Y \end{array}$$

commutes;

(ii) $\varphi$ is surjective and a subset $U \subseteq Y$ is open if and only if $\varphi^{-1}(U)$ is open; such a $\varphi$ is called *submersive*;

(iii) the induced map $\Phi$ in the diagram

$$\begin{array}{ccc} H \times_S X & \xrightarrow{(\omega, p_2)} & X \times_S X \\ & \searrow^{\Phi} & \downarrow \\ & & X \times_Y X \end{array}$$

is surjective;

(iv) for all open $U \subseteq Y$, $\mathcal{O}_Y(U)$ is the difference kernel of the maps

$$\mathcal{O}_X(\varphi^{-1}(U)) \underset{p_2^\sharp}{\overset{\omega^\sharp}{\rightrightarrows}} \mathcal{O}_{H\times X}(H \times_S \varphi^{-1}(U)) \quad .$$

Replacing $H$ by $H \times_S Y$, one can assume that $Y = S$. The pair $(Y, \varphi)$ is called a *universal geometric quotient*, if for any morphism $Y' \to Y$ of $S$-schemes, $X \times_Y Y' \xrightarrow{\varphi'} Y'$ is a geometric quotient, too.

Recall also (c.f. loc. cit., prop. 0.1) that any geometric quotient is a *categorical quotient*. This means that any $H$-morphism from $X$ to an $S$-scheme $Z$ with trivial $H$-action factorizes uniquely over $Y$. In particular, if it exists, a geometric quotient is unique up to an isomorphism. We denote it by $H\backslash X$. Similarly, one defines the geometric quotient $X/H$ if $H$ acts on $X$ from the right.

The map $\varphi : X \to Y$ is called a *principal homogeneous space* with group $H$, if it is flat and surjective, satisfies condition (i), and if the map $\Phi$ in condition (iii) is an isomorphism. This notion is stable under base change $Y' \to Y$.

LEMMA 3.1. *Let $H = (H, \mu)$ be a group over $S$, which is flat and finite over $S$. Let $\varphi : X \to Y$ be a principal homogeneous space over $S$ with group $H$. Then $\varphi$ is a universal geometric quotient.*

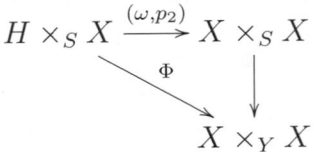

PROOF. We can assume that $Y = S$. Condition (ii) of the definition holds, since $\varphi$ is finite, and consequently a closed map. It remains to check condition (iv). Since $H$ is flat, it is enough to check it for affine open subsets $U \subseteq Y$. Let $B := \mathcal{O}_X(\varphi^{-1}(U))$, $C := \mathcal{O}_Y(U)$, and $P := \mathcal{O}(H \times_Y U)$. Then we must show that $C \to B$ is exactly the difference kernel of the maps $B \underset{p_1^\sharp}{\overset{\omega^\sharp}{\rightrightarrows}} B \otimes_C P$ . Since $B$ is faithfully flat over $C$, tensoring the relevant sequence with $B$ over $C$, it becomes isomorphic to the sequence $C \longrightarrow P \overset{\mu^\sharp}{\underset{p_1^\sharp}{\rightrightarrows}} P \otimes_C P$ . The latter is exact. This proves the lemma. □

For the rest of this section we suppose that $S$ is a connected scheme.

PROPOSITION 3.2. *Let $\mathcal{L}$ be an invertible sheaf over $S$, let $H \subseteq \mathbb{G}_{a,\mathcal{L}}$ be a closed subgroup which is finite and flat over $S$ of rank $n$. Then there exists the universal geometric quotient $\mathbb{G}_{a,\mathcal{L}}/H$. It is a line bundle again. More precisely,*

$$\mathbb{G}_{a,\mathcal{L}}/H \simeq \mathbb{G}_{a,\mathcal{L}^{\otimes n}} .$$

PROOF. First consider the case $S = \operatorname{Spec} R$ and $\mathbb{G}_{a,\mathcal{L}} = \operatorname{Spec} R[X]$. We use the following lemma.

LEMMA 3.3. *Let $H \subseteq \operatorname{Spec} R[X]$ be a closed subscheme which is finite and flat of rank $n$ over $\operatorname{Spec} R$. Then the ideal sheaf of $H$ is generated by a uniquely determined monic polynomial $h \in R[X]$ of degree $n$.*

PROOF. One reduces immediately to the case of a local ring $R$. Then $B := \mathcal{O}(H)$ is finite and free over $R$. Looking at $\bar{B} = B/\mathfrak{m}_R B$ it follows that $\bar{B}$ has a basis of the form $1, \ldots, X^{n-1}$. By Nakayama's lemma it follows, that $B$ has basis $1, X, \ldots, X^{n-1}$. □

Thus we have $H = \operatorname{Spec} R[X]/h$ with a monic polynomial $h$ of degree $n$. Since $H$ is a subgroup, we have $add^\sharp(h) = h(X+Y) = a(X,Y)h(X) + b(X,Y)h(Y)$, where we may assume that the degree of $a$ with respect to $Y$ is less than $n$. Comparing degrees and coefficients, it now follows easily that $h(X+Y) = h(X) + h(Y)$. We claim that the natural projection $\operatorname{Spec} R[X] \overset{p_H}{\to} \operatorname{Spec} R[h]$ is a principal homogeneous space for $H$. Condition (i) holds, since $h$ is additive. So all we have to show is that the canonical map $R[X] \otimes_{R[h]} R[X] \to R[X] \otimes_R R[X]/h$ is bijective. But it is a surjective map of $R[h]$-modules, both free of rank $n^2$. So it is bijective. Moreover, as a group, $\mathbb{G}_{a,R}/H \cong \mathbb{G}_{a,R}$.

In the general case we choose an open affine covering $S = \cup_\alpha S_\alpha$, such that $\mathcal{L}|S_\alpha$ is trivial. Let $S_\alpha = \operatorname{Spec} R_\alpha$. Then

$$\mathbb{G}_{a,\mathcal{L}|S\alpha} = \operatorname{Spec} R_\alpha[X_\alpha] ,$$

and the ideal of $H \times_S S_\alpha$ is generated by a monic polynomial $h_\alpha$. Then, if we have

$$X_\alpha = c_{\alpha\beta} X_\beta$$

over $S_\alpha \cap S_\beta$, we must have

$$h_\alpha(c_{\alpha\beta} X_\beta) = c'_{\alpha\beta} \cdot h_\beta$$

for some $c'_{\alpha\beta} \in \mathcal{O}(S_\alpha \cap S_\beta)^\times$. Looking at the leading coefficients we see that $c'_{\alpha\beta} = c^n_{\alpha\beta}$. Hence the local projections $\operatorname{Spec} R_\alpha[X_\alpha] \to \operatorname{Spec} R_\alpha[h_\alpha]$ glue to a projection

$$p_H : \mathbb{G}_{a,\mathcal{L}} \to \mathbb{G}_{a,\mathcal{L}^{\otimes n}} ,$$

which is a universal geometric quotient by $H$. □

COROLLARY 3.4. *Let $H = V(f) \subseteq H' = V(g)$ be finite and flat subgroups of $\mathbb{G}_{a,S}$ over $S$. Then there exists a polynomial $h$ such that $g = h(f)$. Moreover, $h$ is additive.*

PROOF. Apply Lemma 3.3. to the finite flat group scheme $H'/H$ in $\mathbb{G}_{a,S}/H$. □

## 4. The convergence lemma

In this section we consider additive polynomials over a discrete valuation ring $O$ with quotient field $K$ and multiplicative norm $|\ |$. We also consider additive power series. If such a series commutes with additive polynomials, we can derive strong convergence properties. On the other hand, their coefficients turn out to be integral. More precisely, the following results hold.

LEMMA 4.1. *Let $O$ be a discrete valuation ring with quotient field $K$, let*

$$f = \sum_{i=0}^{n} a_i X^{p^i} \text{ and } g = \sum_{i=0}^{m} b_i X^{p^i}$$

*be additive polynomials in $K[X]$, and let*

$$u = \sum_{i=0}^{\infty} u_i X^{p^i}$$

*be an additive nonzero power series in $K[[X]]$ such that $f(u) = u(g)$. Suppose that either*

> *$m = 0$ and $|b_0| > 1$, or $m > 0$, and additionally in both cases the following conditions hold:*
> (a) *$u$ is strictly convergent, i.e., $\lim |u_i| = 0$, and*
> (b) $|b_m| \geq \max\left(1, |b_i|^{p^{m-i}}, 0 \leq i < m\right)$.

*Then $n \geq m$, and $u$ has infinite radius of convergence.*

PROOF. By Hadamard's criterion, we must show that $\lim |u_j|^{1/p^j} = 0$. We have

$$f(u) = \sum_{j=0}^{\infty} \left( \sum_{i=0}^{n} a_i u_{j-i}^{p^i} \right) X^{p^j}$$

and

$$u(g) = \sum_{j=0}^{\infty} \left( \sum_{i=0}^{m} b_i^{p^{j-i}} u_{j-i} \right) X^{p^j} .$$

Here $u_k := 0$ for $k < 0$.

If $m = 0$, this gives for all $j$

$$\left( b_0^{p^j} - a_0 \right) u_j = \sum_{i=1}^{n} a_i u_{j-i}^{p^i} . \tag{1}$$

Let $\alpha = \max |a_i|$. Thus we have for all $j$ satisfying $|b_0|^{p^j} > |a_0|$ the inequality

$$|u_j| \leq \max_{1 \leq i \leq n} \left( \frac{\alpha}{|b_0|^{p^j}} |u_{j-i}^{p^i}| \right) .$$

Setting $t_j = |u_j|^{1/p^j}$ we find that

$$t_j \leq \max_{1 \leq i \leq n} \left( \frac{\alpha^{1/p^j}}{|b_0|} t_{j-i} \right) .$$

Since $\alpha^{1/p^j}/|b_0| \leq \vartheta < 1$ for $j \gg 0$, we see that the $t$'s tend to 0.

Consider now the case $m > 0$. Comparing coefficients we get

$$|b_m|^{p^{j-m}} |u_{j-m}| \leq \max \left( |b_i^{p^{j-i}} u_{j-i}|, \ 0 \leq i < m; \ |a_i u_{j-i}^{p^i}|, \ 0 \leq i \leq n \right) . \tag{2}$$

Let $u_{j_0}$ be the first coefficient of $u$ not equal to zero. Then $a_0 = b_0^{p^{j_0}}$. In particular, in virtue of (b), we have

$$|a_0| \leq |b_m|^{p^{j_0 - m}} . \tag{3}$$

By a) we get for all sufficiently large $j$

$$|u_{j-i}| < \min \left( \frac{1}{\alpha}, 1 \right) , \ 0 \leq i \leq m ,$$

where $\alpha = \max(|a_i|)$ . For such $j$, using (b) again, we derive from (2)

$$|u_{j-m}| \leq \max \left( |u_{j-i}|, \ 0 \leq i < m; \ \frac{|a_0|}{|b_m|^{p^{j-m}}} \cdot |u_j|; \ |u_{j-i}|^{p^i - 1}, \ 1 \leq i \leq n \right) .$$

But (3) implies that $|a_0|/|b_m|^{p^{j-m}} \leq 1$ for all $j \geq j_0$. Since $|u_{j-i}|^{p^i-1} < |u_{j-i}|$ for $1 \leq i \leq m$, we get the inequality

$$(4) \qquad |u_{j-m}| \leq \max\left(|u_{j-i}|,\ 0 \leq i < m;\ |u_{j-i}|^{p^i-1}, m < i \leq n\right)$$

for all sufficiently large $j$.

If $n \leq m$ we conclude that $|u_{j-m}| \leq |u_{j-m+r}|$ for all $j \gg 0$ and some $r$. Since the $u_i$ tend to zero, $u$ must be a polynomial, which implies $m = n$.

Let $m < n$. Put $s_j = \max(|u_r|, r \geq j)$. Then we have from (4) for $j \gg 0$

$$|u_{j-m}| \leq \max\left(s_{j-m+1},\ s_{j-m-1}^{p^{m+1}-1}, \ldots, s_{j-n}^{p^n-1}\right).$$

The right hand side majorizes the respective expressions for $|u_{j-m+1}| \ldots$ hence we get

$$s_{j-m} \leq \max\left(s_{j-m+1},\ s_{j-m-1}^{p^{m+1}-1}, \ldots, s_{j-n}^{p^n-1}\right).$$

Substituting this inequality for $s_{j-m+1}, s_{j-m+2}, \ldots$ and taking into account that all $s_i$ are $\leq 1$ and $\lim s_i = 0$, the reader may check that even

$$s_{j-m} \leq \max\left(s_{j-m-1}^{p^{m+1}-1}, \ldots, s_{j-n}^{p^n-1}\right).$$

Setting $t_j = s_j^{1/p^j}$, the $t_j$'s satisfy the inequality

$$t_{j-m} \leq \max\left(t_{j-m-1}^{p^m-\frac{1}{p}}, \ldots, t_{j-n}^{p^m-\frac{1}{p^n-p^m}}\right).$$

Since $t_j < 1$ for $j \gg 0$, the $t_j$'s tend to 0 as $j$ tends to infinity. This completes the proof. □

COROLLARY 4.2. *Let $\varphi$ be an endomorphism of $\mathbb{G}_{a,K}$. Suppose $|\partial\varphi| > 1$. Then there exists a unique additive power series $u \in K[[X]]$ with infinite radius of convergence, such that*

(i) $Du(0) = 1$,

(ii) $u^{-1}(\varphi^\sharp(u)) = \partial\varphi \cdot X$.

*Here $u^{-1}$ is the series given by $u^{-1}(u) = X$, and $Du$ is the derivative of $u$.*

PROOF. Let $a_0 = \partial\varphi$, let $v_r$ be an additive series of the form $v_r = a_0 X + a_r X^{p^r} + \ldots$, and let $u_r = X + cX^{p^r}$. Then, modulo $X^{p^{r+1}}$, we have

$$v_{r+1} := u_r^{-1}(v_r(u_r)) = a_0 X + \left(a_r + c(a_0 - a_0^{p^r})\right) X^{p^r}.$$

Since $a_0 \neq a_0^{p^r}$, we may choose $c$ to kill the last term. Composing such polynomials $u_r$, $r > 0$, we obtain a power series transforming $\varphi^\sharp$ to a linear form. This series has infinite radius of convergence by lemma 4.1.

If $u^{-1}(\varphi^\sharp(u)) = v^{-1}(\varphi^\sharp(v)) = a_0X$, the series $w = v^{-1}(u)$ satisfies $w(a_0X) = a_0w(X)$. Thus $w$ is linear. From that uniqueness of $u$ follows. □

COROLLARY 4.3. *Let $O$ be a complete discrete valuation ring, and let $\varphi$ be a quasi-finite endomorphism of $\mathbb{G}_{a,O}$. Then there exists a unique power series $u \in O[[X]]$ with infinite radius of convergence, such that*

(i) $Du(0) = 1$, *and* $u \equiv X \bmod \mathfrak{m}_O[X]$,
(ii) $\psi^\sharp := u^{-1}(\varphi^\sharp(u))$ *gives a finite endomorphism of* $\mathbb{G}_{a,O}$.

PROOF. For any $n$ we apply proposition 2.8 to $\bar{\varphi} = \varphi \bmod \mathfrak{m}_O^n$, so there exists a polynomial $v_n$, such that the leading coefficient of $v_n(\bar{\varphi}^\sharp(v_n^{-1}))$ is a unit and $Dv_n(0) = 1$. By uniqueness, the $v_n$ are congruent $X$ modulo the maximal ideal, and they converge to a series $v$. According to the convergence lemma 4.1., $u := v^{-1}$ has infinite radius of convergence. Since $u \bmod \mathfrak{m}_O^n$ is unique for all $n$, it is unique. □

The convergence lemma admits the following generalization.

LEMMA 4.4. *Let $f$, $g$, $u$ be as in the convergence lemma 4.1., such that*

$$f(u) = u(g) + h$$

*with a polynomial $h = \sum c_i X^{p^i} \in K[X]$. Suppose that either*

$m = 0$ *and* $|b_0| > 1$, *or*
$m > 0$, *$u$ and $g$ are satisfying the conditions* 4.1, (a) *and* (b), *and*
(c) $\max_i(|c_i|/|b_m|^{p^{j-m}}) \leq 1$ *for* $j \gg 0$.

*Then $u$ has infinite radius of convergence; if $u$ is not a polynomial, we have $m < n$.*

PROOF. In case of $m = 0$ we have the equation 4.1, (1), if $j$ is greater than the degree of $h$. In case of $m > 0$ the equation $a_0 = b_0^{p^{j_0}}$ has to be replaced by $a_0 = b_0^{p^{j_0}} + c_{j_0}$. Using (c), one shows that the inequality (4) in the proof of lemma 4.1. continues to hold for $j \gg 0$. As above, this implies the first statement. The second one is proved similarly as above. □

The ring of strict convergent power series is called the *Tate algebra*; it is denoted by $T_{1,K}$ or $T_1$, for short. Let $\overset{\circ}{T_1} := T_1 \cap O[[X]]$ be its canonical lattice.

LEMMA 4.5. *Let* $u \in \overset{\circ}{T_1}$, *such that* $u(0) = 0$ *and* $u \not\equiv 0 \bmod \mathfrak{m}_O \overset{\circ}{T_1}$, *and let* $f \in T_1$. *If* $f(u) \in \overset{\circ}{T_1}$, *then* $f \in \overset{\circ}{T_1}$.

PROOF. Let $t$ be a regular parameter of $\mathfrak{m}_O$. Then, for some $r$, $t^r f \in \overset{\circ}{T_1}$ and $t^r f \not\equiv 0 \bmod \mathfrak{m}_O \overset{\circ}{T_1}$. Since $\deg u \bmod \mathfrak{m}_O \overset{\circ}{T_1} > 0$ we get $t^r f(u) \not\equiv 0 \bmod \mathfrak{m}_O \overset{\circ}{T_1}$, and we conclude that $r \leq 0$. □

COROLLARY 4.6. *Let* $f$, $g$ *be in* $O[X]$, *and let* $0 \neq u \in K[X]$, $u(0) = 0$, *satisfy the following conditions:*

(i) $\deg \bar{f} > 1$ *and* $\deg \bar{g} > 1$, *where* $\bar{f} = f \bmod \mathfrak{m}_O$ *and* $\bar{g} = g \bmod \mathfrak{m}_O$ *;*

(ii) $f(u) = u(g)$.

*Then* $u \in O[X]$ *and* $u \not\equiv 0 \bmod \mathfrak{m}_O[X]$.

PROOF. Let $\lambda \in K$ be an element, such that $u' = \lambda^{-1} u \in O[X]$ and $u' \not\equiv 0 \bmod \mathfrak{m}_O[X]$, and let $f' = \lambda^{-1} f(\lambda X)$. Then $f'(u') = u'(g)$ is in $O[X]$. Because $u' \not\equiv 0 \bmod \mathfrak{m}_O[X]$, we can conclude by lemma 4.5, applying it to $u'$, that $f' \in O[X]$. As $\deg \bar{f} > 1$ we conclude $\lambda \in O$. It follows, that

$$\deg \bar{f}' \cdot \deg \bar{u}' = \deg \bar{u}' \cdot \deg \bar{g}\,.$$

From that we conclude that $\deg \bar{f}' = \deg \bar{g} > 1$. Since $\deg \bar{f} > 1$, we get that $\lambda \in O^{\times}$. □

CHAPTER 2

# Drinfeld Modules

In this chapter we develop the basic theory of Drinfeld modules. Similar as in the theory of elliptic curves, there is a notion of torsion points on a Drinfeld module. A level structure is a uniformization of torsion points. Drinfeld modules with level structure give rise to a module problem. The corresponding moduli space is the main object studied in this work. The analogy to the theory of elliptic curves was settled by many authors in various aspects, see e.g. [**GHR**] or [**GPRV**]

Once and for all we fix a nonsingular projective curve $\mathcal{C}$ over a finite field of constants $\mathbb{F}_q$, where $q$ is a power of $p$, and a closed point $\infty$ on it. Let $A$ be the coordinate ring of the affine subset $\mathcal{C} - \{\infty\}$ of $\mathcal{C}$, and let $k = Q(A)$ be the function field of $\mathcal{C}$. For any closed point of $\mathfrak{p}$ of $\mathcal{C}$ with residue field $k(\mathfrak{p})$, let $v_\mathfrak{p}$ be valuation of $k$ at $\mathfrak{p}$, and let $|\cdot|_\mathfrak{p}$ be the normalized norm. This means

$$|a|_\mathfrak{p} = |k(\mathfrak{p})|^{-v_\mathfrak{p}(a)}.$$

## 1. Analytical definition of Drinfeld modules

In this section we give a preliminary definition of a Drinfeld module, valid over suitable fields. There are two reasons to premise this approach before giving the general definition. Firstly, this construction is quite analogous to the construction of elliptic curves, and the result motivates the definition in the next section. Secondly, using this definition, one can prove that there exist objects as defined later.

Let $k_\infty$ be the completion of $k$ at $\infty$, and let $\ell$ be a complete valuated extension of $k_\infty$. Let $\ell^s$ be its separable closure. A subset $M$ of $\ell^s$ is called *discrete* if $B(0,r) \cap M$ is a finite set for any $r > 0$.

DEFINITION 1.1. A *lattice* over $\ell$ is a finitely generated, projective and discrete $A$-submodule $\Lambda$ of $\mathbb{G}_{a,\ell}(\ell^s)$, which is stable under the action of the Galois group Gal $(\ell^s/\ell)$.

Let $\Lambda_1$ and $\Lambda_2$ be lattices of rank $d$. A *morphism* from $\Lambda_1$ to $\Lambda_2$ is an element $\alpha \in \ell$ such that $\alpha \cdot \Lambda_1 \subseteq \Lambda_2$. Composition is given by multiplication.

Roughly speaking, a Drinfeld module over $\ell$ is the quotient of $\mathbb{G}_{a,\ell}$ by a lattice $\Lambda$. To make this precise, we introduce the following series. We set

$$u_\Lambda = X \cdot \prod_{\substack{\lambda \in \Lambda \\ \lambda \neq 0}} \left(1 - \frac{X}{\lambda}\right) .$$

Since $\Lambda$ is discrete and Galois stable, this product exists as a series with infinite radius of convergence and coefficients in $\ell$.

LEMMA 1.2. *The series $u_\Lambda$ is additive. This means that $u_\Lambda(X+Y) = u_\Lambda(X) + u_\Lambda(Y)$.*

PROOF. Let $\Lambda' \subseteq \Lambda$ be a finite subgroup. Then we claim that the polynomial

$$u := u_{\Lambda'} := X \cdot \prod_{\substack{\lambda \in \Lambda' \\ \lambda \neq 0}} \left(1 - \frac{X}{\lambda}\right)$$

is additive. In fact, since the cardinality of $\Lambda'$ is a power of $p$, both polynomials $u(X+Y)$ and $u(X)+u(Y)$ have leading form $c\cdot(X+Y)^{|\Lambda'|}$, where $c = (\prod_{\lambda\neq 0} \lambda)^{-1}$. Hence $u(X+Y) - u(X) - u(Y)$ is a polynomial of degree less than $|\Lambda'|$ which vanishes on $\Lambda' \times \Lambda'$. Consequently it must vanish identically.

Since $\Lambda$ is the union of its finite subgroups (and a lattice), the coefficients of the polynomials $u_{\Lambda'}$, $\Lambda' \subseteq \Lambda$ finite, converge in $\ell^s$ giving $u_\Lambda$ as a limit, so it is additive, too. □

In particular, $u_\Lambda$ induces a homomorphism of groups

$$u_\Lambda : \mathbb{G}_{a,\ell}(\bar{\ell}) \to \mathbb{G}_{a,\ell}(\bar{\ell})$$

with kernel $\Lambda$. It follows from $u'_\Lambda = 1$ that it is surjective. But, since $\Lambda$ is an $A$-module, the $A$-module structure on $\bar{\ell} = \mathbb{G}_{a,\ell}(\bar{\ell})$ given by the structure morphism pushes down to the quotient and gives a new $A$-module structure on $\bar{\ell}$. It is given by $a \star \bar{x} = u_\Lambda(a \cdot x)$.

PROPOSITION 1.3. *Let $\Lambda$ be a lattice of rank $d$. Then the $A$-module structure on $\mathbb{G}_{a,\ell}(\bar{\ell})$ is algebraic. This means that there exists a homomorphism*

$$e : A \to \mathrm{End}_{\mathbb{F}_q} \mathbb{G}_{a,\ell} ,$$

*such that $a \star \bar{x} = e(a)^\sharp(\bar{x})$ for all $a \in A$ and all $\bar{x} \in \bar{\ell}$.*

*More precisely, $e(a)^\sharp$ is a polynomial of degree $|a|_\infty^d$, and $\partial \circ e$ is just the structure morphism $A \to \ell$.*

PROOF. Let $0 \neq a \in A$, and let $u = u_\Lambda$. By looking at the roots we prove that there is a constant $c \in \ell$ such that

$$u(a \cdot X) = c \cdot \prod_{\bar{y} \in \frac{1}{a}\Lambda/\Lambda} (u(X) - u(y)) \ .$$

In fact, on both sides are series with infinite radius of convergence, whose roots are exactly $\frac{1}{a}\Lambda$, each of them simple. From this the equality above follows. Hence multiplication by $a$ is given by the polynomial

$$e(a)^\sharp = c \cdot \prod_{\bar{y} \in \frac{1}{a}\Lambda/\Lambda} (X - u(y)) \ ,$$

which has degree $\left|\frac{1}{a}\Lambda/\Lambda\right| = |a|_\infty^d$. From $e(a)^\sharp = u(a \cdot u^{-1})$, we see that $\partial(e(a)) = a$. Furthermore for each $a \in \mathbb{F}_q$, the endomorphism $e(a) \in \operatorname{End} \mathbb{G}_{a,\ell}$ is the multiplication by $a$. Hence $e(A)$ takes image in $\operatorname{End}_{\mathbb{F}_q} \mathbb{G}_{a,\ell}$. ☐

Now we can give a precise definition of a Drinfeld module.

DEFINITION 1.4. A *Drinfeld module* of rank $d > 0$ over $\ell$ is a ring homomorphism

$$e : A \longrightarrow \operatorname{End} \mathbb{G}_{a,\ell} \ ,$$

such that

(i) $\partial \circ e$ is the structure morphism of $\ell$.
(ii) $e(a)^\sharp$ is a polynomial of degree $|a|_\infty^d$;

It is automatic then, that for the field of constants $\mathbb{F}_q \subset A$ the image $e(\mathbb{F}_q)$ is given by $\ell$-linear endomorphisms of $\mathbb{G}_{a,\ell}$. Then, clearly, as a homomorphism of rings, $e$ takes image in $\operatorname{End}_{\mathbb{F}_q} \mathbb{G}_{a,\ell}$. Let $e$, $f$ be Drinfeld modules over $\ell$. A *homomorphism* from $e$ to $f$ is an endomorphism $\varphi$ of $\mathbb{G}_{a,\ell}$ satisfying $\varphi \circ e(a) = f(a) \circ \varphi$. By comparing degrees one shows easily that a non zero homomorphism can only exist between Drinfeld modules of the same rank, c.f. 1, 2.7.

The linear action of $A$ on $\ell$ is called the *additive* module.

PROPOSITION 1.5. *The category of Drinfeld modules over $\ell$ of rank $d$ is equivalent to the category of lattices of rank $d$ over $\ell$.*

PROOF. Let $e : A \to \operatorname{End} \mathbb{G}_{a,\ell}$ be a Drinfeld module of rank $d$, and let $a \in A$ with $|a|_\infty > 1$. By 1, 4.2 there exists an additive power series $u$ of infinite radius of convergence such that $Du(0) = 1$ and $u^{-1}(e(a)^\sharp(u)) = a \cdot X$. Now, for any $b \in A$, setting $\tilde{e}_b^\sharp = u^{-1}(e(b)^\sharp(u))$, we have

$$\tilde{e}_b^\sharp(a \cdot X) = a \cdot \tilde{e}_b^\sharp \ .$$

From this we conclude that $\tilde{e}_b^\sharp = b \cdot X$. So the conjugate of $e$ by $u$ is the additive $A$-module over $\ell$. Let $\Lambda$ be the set of roots of $u$. Then by construction $\Lambda$ is a submodule, that is contained in $\ell^s$ and discrete, see the lemma below. Moreover

$$\frac{1}{a}\Lambda/\Lambda \cong (\ker e(a))(\ell^s) \cong (A/a)^d \,.$$

Let $r > 0$, such that the map $B(0,r) \cap \Lambda \to \Lambda/a\Lambda$ is surjective. Then, for any $0 \neq y \in \Lambda$ there exist elements $x$, $z \in \Lambda$ such that $|z|_\infty \leq r$ and $y - z = a \cdot x$. If $|y|_\infty > r$, we have $|x|_\infty < |y|_\infty$. Consequently $\Lambda$ is finitely generated, and therefore is a lattice of rank $d$.

The converse construction was given in 1.3.

Let $\Lambda_1$ and $\Lambda_2$ be lattices of rank $d$, and let $\alpha \in \ell$ be a morphism from $\Lambda_1$ to $\Lambda_2$. By the same argument as in 1.3, there exists a constant $c$ such that

$$u_{\Lambda_2}(\alpha \cdot X) = c \cdot \prod_{\bar{y} \in \alpha^{-1}\Lambda_2/\Lambda_1} (u_{\Lambda_1}(X) - u_{\Lambda_1}(y)) \,.$$

Therefore, multiplication by $\alpha$ pushes down to a homomorphism of Drinfeld modules.

Conversely, any homomorphism of Drinfeld modules produces an endomorphism of the additive module. This is multiplication by a constant and clearly gives a morphism of the corresponding lattices. □

LEMMA 1.6. *Let $u \in \ell[[X]]$ be an additive power series with infinite radius of convergence such that $Du(0) \neq 0$. Then all roots of $u$ in $\bar{\ell}$ are separable, and for any $r > 0$, there is only a finite number of roots of norm less or equal to $r$.*

PROOF. We may assume $r = |\lambda|$ for some $\lambda \in \ell$. Replacing $u$ by $u(\lambda X)$, we may assume $r = 1$. But $u$ is in the Tate algebra $T_{1,\ell}$, c.f. 1, 4.5. By the Weierstrass preparation theorem, see [**BGR**], 5.2, it has a factorization

$$u = h \cdot \omega \,,$$

where $h$ is a monic polynomial with integral coefficients, and $\omega$ is a unit in $T_1$. The roots within the ball with radius 1 are exactly those of $h$, so their number is finite. Taking derivatives, they are separable as well. □

The next proposition shows that in any complete valuated field of rank at least $d$ over $k_\infty$ there exist lattices of rank $d$. Thus there exist Drinfeld modules of rank $d$ over such fields.

PROPOSITION 1.7. *Let $\Lambda$ be an $A$-submodule of $\ell$, which is projective of rank $d$. The following are equivalent:*

(a) *$\Lambda$ is a lattice.*

(b) *The map $\Lambda \otimes_A k_\infty \to \ell$ is injective.*

PROOF. Let $\Lambda'$ be a free submodule of rank $d$ of $\Lambda$. Since $\Lambda/\Lambda'$ is finite, it will be sufficient to prove the result when $\Lambda$ is free. Let $x_1, \ldots, x_d$ be a base.

Assume that $\Lambda \otimes_A k_\infty \to \ell$ is not injective. Then we can find an unbounded sequence

$$y_n = \sum_{i=1}^{d} x_i \otimes \lambda_{ni}$$

in the kernel of that map. As $k_\infty/A$ is compact, we can find $C > 0$ and elements $a_{ni} \in A$ such that $|\lambda_{ni} - a_{ni}|_\infty < C$ holds for all $n, i$. Therefore, if $D > 0$ is such that $|x_i| < D$ for $i = 1, \ldots, d$, we have

$$\begin{aligned} \left| \sum_{i=1}^{d} a_{ni} x_i \right| &= \left| \sum_{i=1}^{d} (\lambda_{ni} + (a_{ni} - \lambda_{ni})) x_i \right| \\ &\leq \sum_{i=1}^{d} |a_{ni} - \lambda_{ni}||x_i| < d \cdot C \cdot D \end{aligned}$$

Therefore $\{\sum_{i=1}^{d} a_{ni} x_i \mid n = 1, 2, \ldots\}$ is an infinite set lying in $\Lambda \cap B(0, d \cdot C \cdot D)$. Thus $\Lambda$ is not a lattice, so (a) implies (b).

Conversely, if $\Lambda$ is not a lattice we can find a sequence

$$z_n = \sum_{i=1}^{d} x_i \otimes b_{ni} \in \Lambda \otimes k$$

such that $\max_i |b_{ni}| = 1$, whose image in $\ell$ converges to 0. Then some subsequence of $(z_n)$ converges to some $0 \neq z \in \Lambda \otimes k_\infty$ which is mapped to 0. $\square$

## 2. The category of Drinfeld modules

Maintaining the notations given at the beginning, we now define Drinfeld modules over arbitrary $A$-schemes $S$. It is a polynomial action of $A$ on a line bundle over $S$. More precisely

DEFINITION 2.1. Let $i : S \to \operatorname{Spec} A$ be a scheme over $\operatorname{Spec} A$. A *Drinfeld module* over $S$ is a pair $\mathbf{E} = (\mathbb{G}_{a,\mathcal{L}}, e)$, where $\mathcal{L}$ is an invertible sheaf over $S$ and $e$ is a homomorphism from $A$ to $\operatorname{End} \mathbb{G}_{a,\mathcal{L}}$ such that

(i) $\partial \circ e = i^\sharp$;

(ii) for any $0 \neq a \in A$, the morphism $e(a)$ is finite, and at any point of $S$, its degree is $> 1$ for some $a \in A$.

The map $i : S \to \operatorname{Spec} A$ is also called the *characteristic* of the Drinfeld module. Mostly we identify it with its image. A point (or a subset) of $\operatorname{Spec} A$ is *away from the characteristic*, if it does not intersect the image of the characteristic.

In the following we shall write $e_a$ instead of $e(a)$. It follows from 1, 2.5 that $e_a$ is flat if $a \neq 0$.

EXAMPLES. (1) Let $\mathcal{C} = \mathbb{P}^1_{\mathbb{F}_q}$, $q = p^r$, and let $\infty$ be the infinite point. Then $A = \mathbb{F}_q[X]$, and a Drinfeld module over an $A$-scheme $S$ is given by specifying an invertible sheaf $\mathcal{L}$ and a finite $q$-linear endomorphism $e_X$ of $\mathbb{G}_{a,\mathcal{L}}$ having linear part $X \cdot \mathrm{id}_{\mathcal{L}^{-1}}$. It is difficult to give other examples explicitly; but there exist Drinfeld modules for any $A$, at least over $k_\infty^s$, c.f. 1.7. For another explicit example, see [**Du**].

(2) Let $\ell$ be a finite extension of $k_\infty$ as considered in the last section. Then the two definitions of Drinfeld modules over $\ell$ coincide.

Let $T$ be a scheme over $S$ and let $\mathbf{E}$ be a Drinfeld module over $S$. Then $\mathbb{G}_{a,\mathcal{L}}(T)$ is canonically an $A$-module, which we denote by $\mathbf{E}(T)$. If $T = \operatorname{Spec} B$ is affine, we denote it by $\mathbf{E}(B)$.

Let $\mathbf{E}$ be a Drinfeld module over the connected scheme $S$. Then, for any $0 \neq a \in A$, $e_a$ has a well defined degree, c.f. the relevant definition after 1, 2.6.

Recall that $|\ |_\infty$ denotes the norm on $k$ belonging to $\infty$.

PROPOSITION 2.2. *Let $S$ be connected. Then the function* $\deg e_a$ *for $a \in A$ defines a norm on $k$ which is equivalent to $|\ |_\infty$. Hence there is a number $d > 0$ such that* $\deg(e_a) = |a|_\infty^d$. *Moreover $d$ is an integer.*

PROOF. The degree behaves multiplicatively and satisfies

$$\deg(e_a + e_b) \leq \max(\deg e_a, \deg e_b).$$

Since $\deg(e_a) \geq 1$ for all $a \neq 0$, it extends to a norm on $k$. But up to equivalence, $|\ |_\infty$ is the only norm such that $|a| \geq 1$ for all $0 \neq a \in A$. It remains to show the last statement. Since the degree is invariant under base change, we may assume that $S = \operatorname{Spec} \ell$ is the spectrum of a field.

By finiteness of the class number of $A$, we can choose an element $t \in A$ such that $t \cdot A = \mathfrak{q}^s$ for a maximal ideal $\mathfrak{q}$ away from the characteristic of $\ell$. Then $e_t^\sharp$ has $\deg(e_t)$ distinct roots in an algebraic closure $\bar{\ell}$ of $\ell$. So we can determine $d$ by counting these roots. They are exactly the elements of $\mathbf{E}(\bar{\ell})$ annihilated by $t$, and they are contained in $\mathbf{E}(\bar{\ell})_{\mathrm{tors}}$, the torsion submodule of $\mathbf{E}(\bar{\ell})$. This torsion module is divisible since for any $0 \neq a \in A$ and any $x \in \mathbf{E}(\bar{\ell})$, there exists a root of $e_a^\sharp - x$ in $\bar{\ell}$.

Denote by $\mathbf{E}(\bar{\ell})_{\mathfrak{q}-\mathrm{tors}}$ the submodule of elements annihilated by some power of $\mathfrak{q}$. Since this is a divisible torsion module over $A_\mathfrak{q}$, we have

$$\mathbf{E}(\bar{\ell})_{\mathfrak{q}-\mathrm{tors}} \cong (k/A_\mathfrak{q})^{d'} ,$$

where $d' = \dim_{A/\mathfrak{q}} \mathrm{Hom}_A(A/\mathfrak{q}, \mathbf{E}(\bar{\ell})_{\mathfrak{q}-\mathrm{tors}})$. So there are precisely $|t|_\mathfrak{q}^{-d'}$ elements in $\mathbf{E}(\bar{\ell})$ which are annihilated by $t$. By the product formula, we have $|t|_\mathfrak{q} = |t|_\infty^{-1}$. So we obtain

$$|t|_\mathfrak{q}^{-d'} = |t|_\infty^{d'} = |t|_\infty^{d} .$$

This shows that $d = d'$ is an integer. □

The number $d$ is called the *rank* of the Drinfeld module $\mathbf{E}$, denoted by $\mathrm{rk}(\mathbf{E})$.

2.3. Let $\mathbf{E}$ be a Drinfeld module over $S$, and let $0 \neq a \in A$ be not a unit. According to 1, 2.8, there exists an automorphism $u$ such that $u \circ e_a \circ u^{-1}$ is standard. Now it follows from 1, 2.7 that $u \circ e_b \circ u^{-1}$ is standard for all $b \in A$. If all $e_a$ are standard, the Drinfeld module is called *standard*. Thus any Drinfeld module is isomorphic to a standard one. If $\mathbf{E}$ is a standard Drinfeld module, then all $e_a$ are $\mathbb{F}_q$-linear. (This is in fact *not* automatic, when $\mathbf{E}$ is *not* a standard Drinfeld module). Hence locally in $S$, $e$ is a homomorphism $A \to R\{\tau_q\}$ for some $A$-algebra $R$.

Let $\mathbf{E} = (\mathbb{G}_{a,\mathcal{L}}, e)$ and $\mathbf{F} = (\mathbb{G}_{a,\mathcal{M}}, f)$ be Drinfeld modules over a scheme $S$. A *homomorphism* from $\mathbf{E}$ to $\mathbf{F}$ is a homomorphism

$$\varphi : \mathbb{G}_{a,\mathcal{L}} \to \mathbb{G}_{a,\mathcal{M}} ,$$

such that $\varphi \circ e_a = f_a \circ \varphi$ for all $a \in A$.

If $\mathbf{E}$ and $\mathbf{F}$ are standard, it follows from 1, 2.7 that, locally in $S$, a homomorphism is zero or standard. Hence locally in $S$, homomorphisms are zero or finite.

A finite homomorphism is called an *isogeny*. If there exists an isogeny between $\mathbf{E}$ and $\mathbf{F}$, these modules are called *isogeneous*. In this case on each connected component of $S$ we have $\mathrm{rk}(\mathbf{E}) = \mathrm{rk}(\mathbf{F})$.

Locally, each isogeny has a well defined *degree*, c.f. Definition after 1, 2.6. The composition of isogenies is an isogeny, and we have

$$\deg(\psi \circ \varphi) = \deg \psi \cdot \deg \varphi .$$

Finally, we note that $\mathrm{Hom}(\mathbf{E}, \mathbf{F})$ is an $A$-module and that $\mathrm{End}(\mathbf{E}) := \mathrm{Hom}(\mathbf{E}, \mathbf{E})$ is an $A$-algebra.

More generally, let $\alpha : S \to T$ be a morphism of $A$-schemes, let $\mathbf{E}$ be a Drinfeld module over $S$, and let $\mathbf{F}$ be a Drinfeld module over $T$. Then we define a *homomorphism* from $\mathbf{E}$ to $\mathbf{F}$ to be a morphism $\varphi : \mathbb{G}_{a,\mathcal{L}_\mathbf{E}} \to \mathbb{G}_{a,\mathcal{L}_\mathbf{F}}$ over $\alpha$ such that $\varphi \times S$ is a homomorphism to $\alpha^*\mathbf{F}$. A similar definition is made for isogenies.

Let $\ell$ be a field over $A$ of characteristic $\mathfrak{p} \neq (0)$, let $\mathbf{E}$ and $\mathbf{F}$ be Drinfeld modules over $\ell$ of rank $d > 0$, and let $\varphi$ be an isogeny from $\mathbf{E}$ to $\mathbf{F}$. Let $r$ be the degree of $k(\mathfrak{p})$ over $\mathbb{F}_q$. Then, if

$$\varphi = \sum_{i=i_0}^{n} a_i \tau_q^i$$

and $a_{i_0} \neq 0$, it follows from $f_a \circ \varphi = \varphi \circ e_a$ that $i_0$ must be a multiple of $r$. The number $h = i_0/r$ is called the *height* of $\varphi$. In particular, let $\mathbf{E} = \mathbf{F}$ and $\varphi = e_t$, where $t \in A$ is a generator of $\mathfrak{p}A_\mathfrak{p}$. The *height* of $\mathbf{E}$ is defined as the height of $e_t$. This definition does not depend on $t$.

If $\mathfrak{p} = (0)$, any isogeny is étale, so its height is defined to be zero.

COROLLARY 2.4. *Let $\ell$ be an algebraically closed field over $A$ of characteristic $\mathfrak{p} \neq (0)$, and let $\mathbf{E}$ be a Drinfeld module of rank $d$ over $\ell$. Then we have*

$$\mathbf{E}(\ell)_{\mathrm{tors}} \cong \bigoplus_{\mathfrak{q} \neq \mathfrak{p}} (k/A_\mathfrak{q})^d \oplus (k/A_\mathfrak{p})^{d-h},$$

*where $h$ is the height of $\mathbf{E}$. Similarly, for $\mathfrak{p} = (0)$ we have*

$$\mathbf{E}(\ell)_{\mathrm{tors}} \cong \bigoplus_{\mathfrak{q}} (k/A_\mathfrak{q})^d .$$

PROOF. For $\mathfrak{q} \neq \mathfrak{p}$ this is in the proof of proposition 2.2. Similarly, if $\mathfrak{p} \neq (0)$, one counts the roots of $e_t^\sharp$, where $t$ generates some power of $\mathfrak{p}$. Note that if $(t) = \mathfrak{p}^s$, then $e_t$ has height $s \cdot h$. Hence there are exactly $|t|_\infty^{d-h}$ elements in $\mathbf{E}(\ell)_{\mathrm{tors}}$, which are annihilated by $t$. □

Let $\mathbf{E}$ and $\mathbf{F}$ be Drinfeld modules over $S$, and let $\varphi : \mathbf{E} \to \mathbf{F}$ be a homomorphism. The difference kernel of the maps $\mathbf{E} \underset{0}{\overset{\varphi}{\rightrightarrows}} \mathbf{F}$ is a closed subgroup of $\mathbf{E}$, which is invariant under the action of $A$. It is called the *kernel* of $\varphi$. If $\varphi$ is an isogeny, its kernel is finite and flat over $S$. In fact, locally in $S$, $\varphi$ identifies with a homomorphism $\varphi : \operatorname{Spec} R[X] \to \operatorname{Spec} R[Y]$. Then its kernel is $\operatorname{Spec} R[X]/\varphi^\sharp(Y)$.

## 3. Drinfeld modules over fields

If the base scheme is the spectrum of a field, there are more precise results concerning homomorphisms of Drinfeld modules. In the next chapter, using deformation theory, we will be able to extend some of these results to the case of arbitrary base schemes as well.

Let $i : A \to \ell$ be an arbitrary field over $A$. To avoid confusion, we say it has *generic characteristic* if it has characteristic $(0)$. Let $\mathbf{E}$ and $\mathbf{F}$ be Drinfeld modules of rank $d > 0$ over $\ell$. Then the structure of the module $\mathrm{Hom}(\mathbf{E}, \mathbf{F})$ is exhibited by looking at the kernel of isogenies.

LEMMA 3.1. *Let*

$$\varphi : \mathbf{E} \longrightarrow \mathbf{F} \text{ and } \psi : \mathbf{E} \longrightarrow \mathbf{G}$$

*be isogenies. If* $\ker \varphi \subseteq \ker \psi$, *then there exists a unique isogeny*

$$\chi : \mathbf{F} \longrightarrow \mathbf{G}$$

*such that* $\psi = \chi \circ \varphi$.

PROOF. By 1, 3.4. there exists a homomorphism

$$\chi : \mathbb{G}_{a,\ell} \longrightarrow \mathbb{G}_{a,\ell}$$

satisfying $\chi \circ \varphi = \psi$. It remains to check that $\chi \circ f_a = g_a \circ \chi$ for $a \in A$. Because $\varphi$ is an epimorphism, this follows from

$$\chi \circ f_a \circ \varphi = g_a \circ \chi \circ \varphi$$

which is clear, because $\varphi$ and $\psi$ are homomorphisms. $\square$

LEMMA 3.2. *Let* $H \subseteq \mathbf{E}$ *be a finite subgroup which is* $A$*-invariant, and let* $\mathfrak{p}$ *be the characteristic of* $\mathbf{E}$. *Then* $H$ *is the kernel of an isogeny if and only if*

(i) $H$ *is reduced in case* $\mathbf{E}$ *has generic characteristic, and*
(ii) $H_{\mathrm{loc}} = \mathrm{Spec}\, \ell[X]/X^{q^{r \cdot s}}$ *for some* $s \in \mathbb{N}$, *where* $H_{\mathrm{loc}}$ *is the component of* $0$, *and* $\mathbb{F}_{q^r}$ *is the residue field at* $\mathfrak{p}$, *otherwise.*

PROOF. $H$ is the kernel of an endomorphism

$$\varphi = \sum_{\nu=n_0}^{n} b_\nu \tau_p^\nu ,$$

where $b_{n_0} \neq 0$ by corollary 3.4. Since $H$ is $A$-invariant, i.e. $\varphi \circ e_a \in (\varphi)$, there exists a homomorphism

$$f : A \to \mathrm{End}(\mathbb{G}_{a,\ell}/H)$$

satisfying $\varphi \circ e_a = f_a \circ \varphi$ for all $a \in A$. It is a Drinfeld module if and only if $\partial(f_a) = \partial(e_a) = i(a)$. Looking at the coefficient of $\tau_p^{n_0}$ we get

$$b_{n_0} i(a)^{p^{n_0}} = \partial(f_a) b_{n_0}.$$

Therefore $\partial(f_a) = i(a)$ is equivalent to $n_0 = 0$ in the case of generic characteristic and to $\overline{a}^{p^{n_0}} = \overline{a}$ for all $\overline{a} \in \mathbb{F}_{q^r}$ in the case of characteristic

$\mathfrak{p}$. This implies $p^{n_0} = q^{rs}$ and $H_{\mathrm{loc}} = \mathrm{Ker}\ \tau_q^{rs} = \mathrm{Spec}\ \ell[X]/X^{q^{rs}}$. The other direction of the proof is immediate. □

EXAMPLE 3.3. Let $\mathbf{E}$ be a Drinfeld module of characteristic $\mathfrak{p} \neq (0)$. Then ker $\tau_q^r$ is $A$-invariant. Thus the quotient is a Drinfeld module. In fact, let $e_a = \sum a_\nu \tau_q^\nu$. Then the quotient is given by $f_a = \sum a_\nu^{q^r} \tau_q^\nu$.

COROLLARY 3.4. *Let $H \subseteq \mathbf{E}$ be the kernel of an isogeny. Then there exists an $a \in A$, $a \neq 0$, such that*

$$\ker\ e_a \supseteq H.$$

*Let $\varphi$ be an isogeny from $\mathbf{E}$ to $\mathbf{F}$. Then there exists an element $a \in A$ and an isogeny $\psi$ from $\mathbf{F}$ to $\mathbf{E}$ such that*

$$\psi \circ \varphi = e_a.$$

*Composing in the opposite way yields $\varphi \circ \psi = f_a$.*

PROOF. Let $\bar{\ell}$ be an algebraic closure of $\ell$. Since $H(\bar{\ell})$ is finite and $A$-invariant, it is annihilated by some $b \neq 0$. Hence $H$ is contained in the kernel of $e_a$ with $a = b$ if the characteristic is generic, and $a = b^{q^{r \cdot s}}$ otherwise. The second statement follows from this and Lemma 3.1. Finally, from the equalities

$$\varphi \circ \psi \circ \varphi = \varphi \circ e_a = f_a \circ \varphi,$$

it follows that $\varphi \circ \psi = f_a$. □

PROPOSITION 3.5. *Let $\mathbf{E}$ and $\mathbf{F}$ be Drinfeld modules of rank $d$ over $\ell$. Then*

(i) $\mathrm{Hom}(\mathbf{E}, \mathbf{F})$ *is a finitely generated projective $A$-module of rank $\leq d^2$.*
(ii) *Let $\mathfrak{q} \in \mathrm{Spec}\, A$ be away from the characteristic of $\ell$. Then the canonical homomorphism*

$$\mathrm{Hom}(\mathbf{E}, \mathbf{F}) \otimes_A \hat{A}_{\mathfrak{q}} \longrightarrow \mathrm{Hom}_{\hat{A}_{\mathfrak{q}}}(\mathbf{E}(\bar{\ell})_{\mathfrak{q}-\mathrm{tors}}, \mathbf{F}(\bar{\ell})_{\mathfrak{q}-\mathrm{tors}})$$

*is injective. Moreover, its cokernel is torsion free.*

PROOF. Fix $a \in A$ with $|a|_\infty > 1$. First observe that $M = \mathrm{Hom}(\mathbf{E}, \mathbf{F})$ is a torsion free $A$-module. Next we show that it is finitely generated. Endow $A$ with the filtration $A_n := \{a \in A \mid \deg e_a \leq q^n\}$. Then $M$ is a filtered $A$-module with filtration $M_n = \{\varphi \in M \mid \deg \varphi \leq q^n\}$. Let $c$ be the leading coefficient of some $\varphi \in M$ of degree $q^n$, and let $\alpha$ and $\beta$ be the leading coefficients of $e_a$ and $f_a$. Comparing leading coefficients of the equality $f_a \circ \varphi = \varphi \circ e_a$, we conclude

$$c \cdot \alpha^{q^n} = \beta \cdot c^{|a|_\infty^d}.$$

Thus $M_n/M_{n-1}$ has cardinality less than or equal to $|a|_\infty^d - 1$. Since multiplication by $a$ (element of degree $|a|_\infty$) is injective on $\operatorname{gr} M$, it follows easily that $\operatorname{gr} M/a \cdot \operatorname{gr} M$ is finite. Hence $\operatorname{gr} M$ is finitely generated, thus $M$ is finitely generated as well.

Next we prove (ii). Let $M'$ denote the right hand side in (ii), and let $a \in A$ be an element that generates a $\mathfrak{q}$-primary ideal. The map $M \to M'$ is clearly injective. If some isogeny $\varphi$ is mapped to $a \cdot M'$, its kernel contains the kernel of $e_a$. By Lemma 3.1, we have $\varphi = \psi \circ e_a$ for some $\psi$. We conclude that

$$M \cap a^n M' = a^n M$$

for all $n$. Hence the induced map $\hat{M} \to M'$ is injective. Since $M$ is finitely generated, this is the first part of the assertion. The last equation shows also that $a$ acts injectively on the cokernel, so it has no torsion.

Now the second part of (i) follows from $M' \cong M_d(\hat{A}_{\mathfrak{q}})$. □

## 4. Level structures

Similar to the theory of elliptic curves, the subscheme of division points of a Drinfeld module plays an important role. A uniformization of this subscheme is called a level structure. Adding suitable level structures will make Drinfeld modules rigid in the sense that these objects have no automorphism except the identity.

Let $\mathbf{E}$ be a Drinfeld module of rank $d > 0$ over a scheme $S$ and let $I \subseteq A$ be an ideal. The contravariant functor on $S$-schemes

$$T \longmapsto \{x \in \mathbf{E}(T) : Ix = 0\} = \operatorname{Hom}_A(A/I, \mathbf{E}(T))$$

is represented by a closed subscheme $\mathbf{E}[I]$ of $\mathbf{E}$.

In fact, if $I$ is generated by $a_1, \ldots, a_n$, we have

$$\mathbf{E}[I] = \ker(\mathbf{E} \overset{e_{a_1}, \ldots, e_{a_n}}{\longrightarrow} \mathbf{E} \times_S \ldots \times_S \mathbf{E})\,.$$

Hence locally ($S = \operatorname{Spec} R$)

$$\mathbf{E}[I] = \operatorname{Spec} R[X]/(e_{a_1}^\sharp, \ldots, e_{a_n}^\sharp).$$

Of course, any $I$ can be generated by at most 2 elements. This subscheme is called the scheme of $I$-*division points* of $\mathbf{E}$. It has the following basic properties

PROPOSITION 4.1. *Let* $\mathbf{E}$ *be a Drinfeld module of rank* $d$, *and let* $(0) \neq I$ *be a proper ideal of* $A$.

(i) *There exists an ideal* $J \subseteq A$ *such that* $I \cdot J$ *is principal and* $I + J = A$.

(ii) *Let $J$ be an ideal such that $I+J=A$. Then there is a canonical isomorphism*

$$\mathbf{E}[I\cdot J]\simeq \mathbf{E}[I]\times_S \mathbf{E}[J] \quad .$$

(iii) *$\mathbf{E}[I]$ is a finite and flat group scheme of rank $|A/I|^d$ over $S$.*

(iv) *If $I$ is away from the characteristic of $\mathbf{E}$, then $\mathbf{E}[I]$ is étale over $S$.*

(v) *Let $T$ be a $S$-scheme. Then*

$$\mathbf{E}[I]\times_S T\cong(\mathbf{E}\times_S T)[I].$$

PROOF. (i). Let

$$I=\prod_{i=1}^{r}\mathfrak{p}_i^{s_i},\qquad s_i>0.$$

Choose $a\in A$ such that $v_{\mathfrak{p}_i}(a)=s_i$, $1\leq i\leq r$. Then the ideal $J$ given by $I\cdot J=(a)$ has the required properties.

(ii). We have $A/IJ\cong A/I\times A/J$, by the chinese remainder theorem. We conclude that

$$\mathbf{E}[I\cdot J](T)\cong\mathbf{E}[I](T)\times\mathbf{E}[J](T)$$

for any $S$-scheme $T$. This implies (ii).

(iii). We may assume that $S=\operatorname{Spec}R$ is affine. Then the result is immediate if $I$ is principal. Choose $a$ and $J$ as in (i). Using (ii), there is a canonical projection $\mathbf{E}[(a)]\to\mathbf{E}[I]$. Hence, as an $R$-module, $\mathcal{O}(\mathbf{E}[I])$ is a direct summand of $\mathcal{O}(\mathbf{E}[(a)])$, so it is finite and projective.

To compute its rank, we may assume $S=\operatorname{Spec}\ell$ is the spectrum of an algebraically closed field. Let $\mathfrak{p}$ be its characteristic and let $\mathfrak{q}$ be a maximal ideal of $A$ away from $\mathfrak{p}$. Then we have by 2.4,

$$\mathbf{E}[\mathfrak{q}^n](\ell)\simeq(\mathfrak{q}^{-n}A_{\mathfrak{q}}/A_{\mathfrak{q}})^d,$$

hence

$$\dim_\ell\mathcal{O}(\mathbf{E}[\mathfrak{q}^n])\geq|A/\mathfrak{q}^n|^d.$$

Choose $a\in A$ with $v_{\mathfrak{q}}(a)=n$ and $a\notin\mathfrak{p}$. Then $\mathbf{E}[(a)]=\operatorname{Spec}\ell[X]/e_a^\sharp$. Applying the inequality to each primary component of $(a)$ and using (ii), we get equality for all $\mathfrak{q}\neq\mathfrak{p}$. If $\mathfrak{p}\neq(0)$, we now choose $a\in A$ such that $v_{\mathfrak{p}}(a)=n$ and conclude as before.

(iv). Again, let $S=\operatorname{Spec}R$ be affine, and let $a$, $b$ be generators of $I$. Then, over $\operatorname{Spec}R_a$, the assertion is true, since $\mathbf{E}[I]$ is a closed subscheme of $\mathbf{E}[(a)]=V(e_a^\sharp)$, which is flat over $R_a$, and the latter is étale, since $a$ is a unit in $R_a$. The same holds for $\operatorname{Spec}R_b$.

(v). Both $T$-schemes represent the same functor. □

It follows from this proposition and 1, 3.3 that $\mathbf{E}[I]$ is a relative Cartier divisor of $\mathbf{E}$, i.e. a closed subscheme which is flat over $S$ and whose ideal sheaf in $\mathcal{O}_{\mathbf{E}}$ is invertible.

On the other hand, let $x \in \mathbf{E}(S)$ be a section. Then $x(S)$ is a relative Cartier divisor which we denote by $x$ again.

DEFINITION 4.2. Let $\mathbf{E}$ be a Drinfeld module of rank $d > 0$ and let $0 \neq I \subset A$ be an ideal. A *level $I$ structure* on $\mathbf{E}$ is a homomorphism of $A$-modules

$$\iota : (I^{-1}/A)^d \to \mathbf{E}(S)$$

which induces an equality of divisors

$$\mathbf{E}[\mathfrak{p}] = \sum_{\alpha \in (\mathfrak{p}^{-1}/A)^d} \iota(\alpha)$$

for each prime $\mathfrak{p} \in V(I)$. The pair $(\mathbf{E}, \iota)$ is called a *Drinfeld module with level $I$ structure.*

A *total level structure* (or level (0) structure) on $\mathbf{E}$ is a homomorphism $(k/A)^d \to \mathbf{E}(S)$ whose restriction to $(I^{-1}/A)^d$ is a level $I$ structure for each proper ideal $I$.

Note that the notion of a level $I$ structure is invariant under base change.

EXAMPLES 4.3. (1) Let $\ell$ be an algebraically closed field, and let $\mathbf{E}$ be a Drinfeld module over $\ell$. In this case a level $I$ structure is an epimorphism from $(I^{-1}/A)^d$ onto $\mathbf{E}[I](\ell)$. Hence there exists a level $I$ structure on $\mathbf{E}$.

(2) If $V(I)$ is away from the characteristic of $S$, a level $I$ structure is just an isomorphism

$$S \times (I^{-1}/A)^d \xrightarrow{\sim} \mathbf{E}[I] \ .$$

In fact, a level $I$ structure induces a morphism $\iota : S \times (I^{-1}/A)^d \to \mathbf{E}[I]$. To show that $\iota$ is an isomorphism, we argue as follows:
By 4.1.(iv) both schemes are finite and étale over $S$, hence $\iota$ is finite étale. Since by (1), $\iota$ is an isomorphism, when $S$ is an algebraically closed field, it is an isomorphism in general.

The converse is obvious. In particular, after étale base change, any Drinfeld module has a level $I$ structure.

DEFINITION 4.4. Let $(\mathbf{E}, \iota)$ and $(\mathbf{F}, \iota')$ be Drinfeld modules with level $I$ structures. A *morphism* from $(\mathbf{E}, \iota)$ to $(\mathbf{F}, \iota')$ is an isomorphism $\varphi : \mathbf{E} \to \mathbf{F}$ such that

$$\varphi(S) \circ \iota = \iota'.$$

## 5. Modular manifolds

Let $I \subseteq A$ be an ideal not equal to $(0)$. We now consider the functor $\mathbb{M}_I^d$, which to a scheme $S$ over $A$ associates the set of isomorphy classes of Drinfeld modules of rank $d$ and level $I$ over $S$.

PROPOSITION 5.1. *If $V(I)$ contains at least two closed points of* $\operatorname{Spec} A$, *the functor $\mathbb{M}_I^d$ is represented by a scheme $M_I^d$ which is affine and of finite type over $A$.*

PROOF. Let $\mathfrak{p}$ and $\mathfrak{q}$ be two different closed points in $V(I)$. We set

$$B := \mathcal{O}(\operatorname{Spec} A - \{\mathfrak{p}\}) \text{ and } B' := \mathcal{O}(\operatorname{Spec} A - \{\mathfrak{q}\})$$

Then $B$, $B'$ are the rings of the affine sets $\operatorname{Spec} A - \{\mathfrak{p}\}$, $\operatorname{Spec} A - \{\mathfrak{q}\}$ respectively.

*Step 1.* Let $(\mathbf{E}, \iota)$ be a Drinfeld module over a $B$-scheme $S$. Then the underlying line bundle $\mathcal{L}$ of $\mathbf{E}$ is trivial. In fact, $\iota$ induces a $\mathfrak{p}$-level structure which is an isomorphism from $(\mathfrak{p}^{-1}/A)^d \times S$ to $\mathbf{E}[\mathfrak{p}]$. For any $x \in (\mathfrak{p}^{-1}/A)^d - \{0\}$, the section of $\mathcal{L}_{\mathbf{E}}$ belonging to $\iota(x)$ does not vanish in any point of $S$.

*Step 2.* The restriction of $\mathbb{M}_I^d$ to the schemes over $\operatorname{Spec} B$ is represented by an affine scheme of finite type over $\operatorname{Spec} B$. Let

$$A = \mathbb{F}_p[\alpha_1, \ldots, \alpha_n]/(f_1, \ldots, f_r)\,.$$

We denote the image of $\alpha_\nu$ again by $\alpha_\nu$. Let

$$d_\nu := d \cdot \log_q(|\alpha_\nu|_\infty)\,.$$

Then we let

$\tilde{R} := B[\alpha_{\nu,i_\nu}, \alpha_{\nu,d_\nu}^{-1}, \beta_x]$, where $1 \le \nu \le n,\ 1 \le i_\nu \le d_\nu,\ x \in (I^{-1}/A)^d\,.$

Now we set

$$\tilde{e}_\nu := \alpha_\nu \tau^0 + \sum_{i_\nu=1}^{d_\nu} \alpha_{\nu,i_\nu} \tau_q^{i_\nu} \in \operatorname{End}(\mathbb{G}_{a,\tilde{R}})\,,$$

and define a section $\tilde{\iota}(x)$ of $\mathbb{G}_{a,\tilde{R}}$ by

$$\tilde{\iota}(x)^\sharp(X) = \beta_x\,.$$

Finally, we fix an element $x_0 \in (\mathfrak{p}^{-1}/A)^d - \{0\}$. Then our functor is represented by $R := \operatorname{Spec} \tilde{R}/\mathfrak{r}$, where $\mathfrak{r}$ is the ideal generated by following relations.

(i) $\tilde{e}_\nu \circ \tilde{e}_\mu = \tilde{e}_\mu \circ \tilde{e}_\nu$, $1 \le \mu, \nu \le n$;

(ii) $f_\rho(\tilde{e}_1, ..., \tilde{e}_n) = 0$, $1 \leq \rho \leq r$;
(iii) $\tilde{\iota}(x+y) = \tilde{\iota}(x) + \tilde{\iota}(y)$, $x, y \in (I^{-1}/A)^d$;
(iv) $\tilde{\iota}(\alpha_\nu \cdot x) = \tilde{e}_\nu \circ \tilde{\iota}(x)$, $1 \leq \nu \leq n, x \in (I^{-1}/A)^d$;
(v) $\sum_{x \in (\mathfrak{m}^{-1}/A)^d} \tilde{\iota}(x) = \mathbf{E}[\mathfrak{m}]$, $\mathfrak{m} \in V(I)$, where we have identified $\tilde{\iota}(x)$ with its associated divisor, and $\mathbf{E}$ is defined below;
(vi) $\beta_{x_0} = 1$.

To show this, note that modulo the relations (i), the relations (ii) are well defined. By (i) and (ii) we have defined a Drinfeld module $\mathbf{E}$ over $R$ which is provided with a level $I$ structure $\iota$ according to (iii)-(v). Let $(\mathbf{F}, \kappa)$ be a Drinfeld module over the $B$-scheme $S$. We may assume that $S = \operatorname{Spec} C$ is affine. Then there exists a unique standard module $(\mathbf{F}', \kappa')$ isomorphic to $(\mathbf{F}, \kappa)$ such that $\kappa'(x_0)^\sharp(X) = 1$. It is clear that there exists a unique map from $R$ to $C$ such that $(\mathbf{E}, \iota)$ specializes to $(\mathbf{F}', \kappa')$ showing the result.

*Step 3.* In the same way, starting with $B'$ and distinguishing an element $y_0 \in (\mathfrak{q}^{-1}/A)^d$, we obtain a universal Drinfeld module $(\mathbf{E}', \iota')$ of level $I$ over an $B'$-algebra $R'$. Due to universality, there is a canonical isomorphism over $B \otimes_A B'$

$$\eta : R \otimes_A B' \to B \otimes_A R' \ .$$

By glueing $\operatorname{Spec} R$ and $\operatorname{Spec} R'$ along the morphism given by $\eta$, we obtain a scheme $M_I^d$ that represents $\mathbb{M}_I^d$. Since it is affine over Spec $A$, it is affine. □

REMARKS 5.2. (1) The bundle of the universal Drinfeld module is trivial over $\operatorname{Spec} R$ and $\operatorname{Spec} R'$, where it is generated by $\iota(x_0)$ and $\iota(y_0)$, respectively. The glueing map is given by

$$\iota(y_0) = \frac{1}{\beta_{y_0}} \iota(x_0) \, .$$

(2) *Functoriality.* For $J \subseteq I$, $I$ as above, there is a canonical morphism $M_J^d$ to $M_I^d$ corresponding to restriction of level structures. All schemes being affine, there is an affine scheme

$$M^d := \varprojlim M_I^d = \operatorname{Spec} \varinjlim \mathcal{O}(M_I^d)$$

This scheme represents the functor which associates to $S$ the Drinfeld modules of rank $d$ endowed with a total level structure.

EXAMPLE 5.3. If $I = \mathfrak{p}^n$ is a primary ideal, or if $I = A$ (case of no level structure), the functor $\mathbb{M}_I^d$ is not representable. To see this, consider the following example.

Let $p \neq 2$, let $A = \mathbb{F}_p[T]$, and let $I = (T)$. Let

$$R = R' := A[U, V]/(T, V^{p+1} - 2U^{p+1}) \ .$$

Then we have

$$R_U = R'_U = \mathbb{F}_p[U, U^{-1}, W]/(W^{p+1} - 2) ,$$

where $W = VU^{-1}$. Let $S$ be the scheme obtained by glueing $\operatorname{Spec} R$ and $\operatorname{Spec} R'$ along $\operatorname{Spec} R_U$ by means of the identity. Let $\mathcal{L}$ be the invertible sheaf obtained by glueing the trivial bundles over $\operatorname{Spec} R$ and $\operatorname{Spec} R'$ by means of multiplication by $W$. It is easy to verify that $\mathcal{L}$ is not trivial, but $\mathcal{L}^{p^2-1}$ is. Let $s$, $s'$ be trivializing sections of $\mathcal{L}$ over $\operatorname{Spec} R$, $\operatorname{Spec} R'$, respectively, such that $s' = W \cdot s$ over the intersection. Consider the Drinfeld modules $\mathbf{E}$ and $\mathbf{F}$ of rank 2 with bundles $\mathcal{L}$ and $\mathcal{O}_S$, respectively, given by $\tilde{e}_T = s^{p^2-1}$ and $\tilde{f}_T = (\tilde{f}_T)_2 = 1$, see 1, 2.3 for notations. On both Drinfeld modules, $\iota = 0$ is a level $\mathfrak{p}$ structure. The restrictions of $\mathbf{E}$ and $\mathbf{F}$ to $\operatorname{Spec} R$ and $\operatorname{Spec} R'$ are isomorphic, but $\mathbf{E}$ and $\mathbf{F}$ are not. Hence the presheaf $S' \mapsto \mathbb{M}^2_{\mathfrak{p}}(S')$, $S' \subseteq S$ open, is not a sheaf. So $\mathbb{M}^2_{\mathfrak{p}}$ is not representable.

If $|V(I)| \geq 2$, the only automorphism of a Drinfeld module of level $I$ is the identity. In fact, assuming standard form, any automorphism is linear. Since the level structure is not zero at any point of the base, it must be the identity.

## 6. Pseudo-Drinfeld modules

For the construction of the boundary of our moduli problem in chapter 5 we need the following concept of a pseudo-Drinfeld module. The idea here is to be able to speak about Drinfeld modules on arbitrary ringed spaces over the curve $\operatorname{Spec}(A)$, which turn out to be actually schemes in our situation only later on.

So, we start with a ringed space

$$(X, \mathcal{O}_X) \to \operatorname{Spec}(A)$$

in the sense of [**EGA I**], $\mathcal{L}$ denotes an invertible sheaf of modules over $(X, \mathcal{O}_X)$, $\mathcal{S}_{\mathcal{O}_X}(\mathcal{L}^{-1}) = \bigoplus_{n \geq 0} (\mathcal{L}^{-1})^{\oplus n}$ is the symmetric algebra of $\mathcal{L}^{-1}$ over $X$, which is a sheaf of commutative algebras and is locally on $X$ isomorphic to $\bigoplus_{n \geq 0} (\mathcal{O}_U)^{\oplus n} \cong \mathcal{O}_U[T]$ for $U$ open in $X$, $\mathcal{L}|U \cong \mathcal{O}_U$.

Obviously, it makes sense to speak about the additive group, associated to $\mathcal{L}^{-1}$, over the ringed space $(X, \mathcal{O}_X)$. This is given by the functor

$$U \mapsto (\mathcal{L}^{-1}(U); +)$$

into the category of abelian groups.

Alternatively, we remark, that $\mathcal{S}_{\mathcal{O}_X}(\mathcal{L}^{-1})$ has the structure of a sheaf of bialgebras, where the Hopf-algebra structure is given as follows: We consider the diagonal embedding

$$\begin{aligned} \Delta : \mathcal{L}^{-1} &\to \mathcal{L}^{-1} \oplus \mathcal{L}^{-1} \\ l &\mapsto (l, l) \end{aligned}$$

By the universality of the symmetric algebra construction, this induces canonically a homomorphism of commutative algebras

$$\tilde{\Delta} : \mathcal{S}_{\mathcal{O}_X}(\mathcal{L}^{-1}) \to \mathcal{S}_{\mathcal{O}_X}(\mathcal{L}^{-1} \oplus \mathcal{L}^{-1})$$

Furthermore, we have the canonical isomorphism of sheaves of $\mathcal{O}_X$-algebras

$$\mathcal{S}_{\mathcal{O}_X}(\mathcal{L}^{-1} \oplus \mathcal{L}^{-1}) \xrightarrow{\sim} \mathcal{S}_{\mathcal{O}_X}(\mathcal{L}^{-1}) \otimes \mathcal{S}_{\mathcal{O}_X}(\mathcal{L}^{-1})$$

Combining this, we obtain on $\mathcal{S}_{\mathcal{O}_X}(\mathcal{L}^{-1})$ additionally the structure of a sheaf of $\mathcal{O}_X$-coalgebras, again called $\Delta$ by abuse of notation

$$\Delta : \mathcal{S}_{\mathcal{O}_X}(\mathcal{L}^{-1}) \longrightarrow \mathcal{S}_{\mathcal{O}_X}(\mathcal{L}^{-1}) \otimes_{\mathcal{O}_X} \mathcal{S}_{\mathcal{O}_X}(\mathcal{L}^{-1}),$$

such that in toto $\mathcal{S}_{\mathcal{O}_X}(\mathcal{L}^{-1})$ is a sheaf of bialgebras on the ringed space $(X, \mathcal{O}_X)$.

DEFINITION 6.1. $\mathrm{End}^{\mathrm{bdl}}(\mathcal{S}_{\mathcal{O}_X}(\mathcal{L}^{-1}))$ denotes the ring of all endomorphisms of the sheaf of bialgebras $(\mathcal{S}_{\mathcal{O}_X}(\mathcal{L}^{-1}); \Delta)$.

DEFINITION 6.2. A pseudo-Drinfeld module of rank $d$ over the ringed space $(X, \mathcal{O}_X)$ is a homomorphisms

$$\varphi : A \to \mathrm{End}^{\mathrm{bdl}}(\mathcal{S}_{\mathcal{O}_X}(\mathcal{L}^{-1})),$$

such that the induced homomorphisms

$$\varphi_U : A \to \mathrm{End}(\mathcal{S}_{\mathcal{O}_X(U)}(\mathcal{L}^{-1}(U)))$$

are Drinfeld-modules over Spec $(\mathcal{O}_X(U))$ of rank $d$ for all open $U \subset X$ in the sense of Definition 2.1. of chapter 2..

REMARK 6.3. Here, by definition the homomorphism into the zero ring is a Drinfeld module of rank $d$.

Similarly, a level I structure on a pseudo-Drinfeld module $\varphi : A \to \mathrm{End}^{\mathrm{bdl}}(\mathcal{S}_{\mathcal{O}_X}(\mathcal{L}^{-1}))$ is a homomorphism of $A$-modules

$$\iota^{\sharp} : (I^{-1}/A)^d \to \Gamma(X, \mathcal{L}),$$

which induces a level I structure on Spec $\mathcal{O}_X(U)$ in the sense of Definition 4.2. of chapter 2. for each open $U \subset X$.

We assume again now, that $|V(I)| \geq 2$.

PROPOSITION 6.4. *Assume* $|V(I)| \geq 2$. *Then there is a canonical bijection of the set of pseudo-Drinfeld modules of rank d and level I structure over* $\mathcal{O}_X$*, up to isomorphism, with* $\mathrm{Hom}_{A\text{-alg}}(\mathcal{O}(M_I^d), \mathcal{O}_X(X))$.

REMARK. Hence a pseudo-Drinfeld module of level I is the same as a Drinfeld module of level I over the global ring of the structure sheaf.

PROOF of Proposition 6.4.: By [**EGA I**], Prop. (2.3.2) the set of morphisms $\mathrm{Hom}_A((X, \mathcal{O}_X), M_I^d)$ in the category of ringed spaces, are in bijective correspondence with $\mathrm{Hom}_A(\mathcal{O}(M_I^d), \mathcal{O}_X(X))$, as $M_I^d$ is an affine scheme over $A$. As $M_I^d$ is a fine moduli scheme for Drinfeld modules with level I structure, it follows immediately, that this describes exactly the set of isomorphism classes of pseudo-Drinfeld modules. □

CHAPTER 3

# Deformation Theory

In this chapter we investigate the moduli spaces $M_I^d$ in more detail. The main result is that these are regular spaces of dimension $d$ and smooth over $\operatorname{Spec} A$ outside the characteristic. To achieve this result, we consider deformation theory of Drinfeld modules à la Schlessinger, [**S**]. We proceed in three steps: 1. Drinfeld modules; 2. isogenies; 3. level structures. Our approach is a generalization of Laumon's, [**La**]. In the last section, we introduce a group action on the moduli spaces that will be important later on.

## 1. Deformations of Drinfeld modules

Let $A \xrightarrow{i} O$ be a complete noetherian local $A$-algebra with residue field $\ell$. Let $\mathcal{C}_O$ be the category of local artinian $O$-algebras with residue field $\ell$. Let $\hat{\mathcal{C}}_O$ be the category of noetherian complete local $O$-algebras with residue field $\ell$.

DEFINITION 1.1. Let $\mathbf{E}_0$ be a Drinfeld module of rank $d > 0$ over $\ell$, and let $B$ be an algebra in $\mathcal{C}_O$. A *deformation* of $\mathbf{E}_0$ over $B$ is a Drinfeld module over $\operatorname{Spec} B$ together with an isomorphism between $\mathbf{E}$ mod $\mathfrak{m}_B$ and $\mathbf{E}_0$.

Let $\mathbf{E}$ and $\mathbf{E}'$ be deformations of $\mathbf{E}_0$ over $B$. They are *isomorphic*, if there is an isomorphism $\mathbf{E} \to \mathbf{E}'$, which specializes mod $\mathfrak{m}_B$ to the identity of $\mathbf{E}_0$.

Let $B \xrightarrow{\varphi} B'$ be a morphism of $O$-algebras in $\mathcal{C}_O$, and let $\mathbf{E}$ be a deformation over $B$. Then $\varphi_*\mathbf{E} := \mathbf{E} \otimes_B B'$ is a deformation over $B'$. So we have a covariant functor $\operatorname{Def}_{\mathbf{E}_0}$ on $\mathcal{C}_O$ which to any $O$-algebra $B$ in $\mathcal{C}_O$ associates the set of deformations of $\mathbf{E}_0$ over $B$, up to isomorphism.

According to 2, 2.3, each deformation is isomorphic to a standard Drinfeld module. Isomorphisms between standard deformations are linear, i.e. of the form $(1+m)\tau^0$, $m \in \mathfrak{m}_B$. Let $(1+m)\tau^0$ be an automorphism of a standard deformation $\mathbf{E}$, and let $m \equiv 0 \bmod \mathfrak{m}_B^s$. Then we have mod $\mathfrak{m}_B^{s+1}$

$$e_a \circ (1+m)\tau^0 = e_a + m \cdot \partial e_a \cdot \tau^0 = (1+m)\tau^0 \circ e_a = e_a + me_a$$

for all $a \in A$. This shows that $m \equiv 0 \bmod \mathfrak{m}^{s+1}$, so $m = 0$. Hence there are no automorphisms unless the identity automorphism, so isomorphisms are unique.

PROPOSITION 1.2. *The functor* $D = \mathrm{Def}_{\mathbf{E}_0}$ *is pro-represented by the smooth* $O$*-algebra* $R_0 = O[[T_1, \ldots, T_{d-1}]]$.

PROOF. As usual, we have to check the conditions of [**S**], (2.5).
(1) We will show that the functor $D$ is homogeneous. This means, it commutes with certain fibre products. More precisely, consider a diagram in $\mathcal{C}_O$

$$\begin{array}{ccc} & & B_2 \\ & & \downarrow{\scriptstyle \varphi_2} \\ B_1 & \xrightarrow{\varphi_1} & B \end{array}$$

with $\varphi_1$ surjective. Let $(\mathbf{E}_1, e^{(1)})$ resp. $(\mathbf{E}_2, e^{(2)})$ be deformations over $B_1, B_2$ respectively, such that $\varphi_{1*}\mathbf{E}_1 \tilde{\rightarrow} \varphi_{2*}\mathbf{E}_2$. This isomorphism can be lifted to an isomorphism from $\mathbf{E}_1$ to some deformation $\mathbf{E}_1'$ over $B_1$. So we may assume that $\varphi_{1*}\mathbf{E}_1 = \varphi_{2*}\mathbf{E}_2$. Then $e^{(1)}$ and $e^{(2)}$ are the components of a deformation over $B_1 \times_B B_2$. Similar reasoning for isomorphisms. Thus the map

$$D(B_1 \times_B B_2) \to D(B_1) \times_{D(B)} D(B_2)$$

is bijective. This is the condition $(\mathrm{H}_4)$ of [**S**], (2.11).

(2) Next we will prove that $D$ is smooth. Let

$$0 \to (\varepsilon) \to B \to \bar{B} \to 0$$

be an extension such that $\mathfrak{m}_B \cdot \varepsilon = 0$ (such an extension is called *small*), and let $\bar{\mathbf{E}}$ be a deformation over $\bar{B}$. Then we must construct a lifting of $\bar{\mathbf{E}}$ to a deformation over $B$. Let $a \mapsto e_a$ be an $\mathbb{F}_q$-linear lifting of $a \mapsto \bar{e}_a$ such that $\partial(e_a) = i(a) = a \cdot 1_B$. It is a lifting of $\bar{\mathbf{E}}$ if and only if $e_a \circ e_b = e_{ab}$. Consider the map

$$t : (a, b) \mapsto e_{ab} - e_a \circ e_b \ .$$

It is a Hochschild 2-cocycle with values in $\varepsilon \cdot B\{\tau\} \cdot \tau$, where $\tau = \tau_q$. Note that the $A$-bimodule $\varepsilon \cdot B\{\tau\}$ is isomorphic to $\mathfrak{t}_{\mathbf{E}_0} := \ell\{\tau\}$ with multiplication $a \cdot e \cdot b = i(a) \cdot e \circ e_b$. Thus $t$ is a cocycle with values in $\mathfrak{t}_{\mathbf{E}_0}^+ := \mathfrak{t}_{\mathbf{E}_0} \cdot \tau$.

In fact, the cocycle condition is

$$a \cdot t(b, c) - t(ab, c) + t(a, bc) - t(a, b) \cdot c = 0$$

for all $a, b, c$. It is verified just by inserting the definition. Hence $t$ determines a class in $\mathrm{Ext}^2_{A \otimes A}(A, \mathfrak{t}^+_{\mathbf{E}_0})$, where the tensor product is taken over $\mathbb{F}_q$. By the following lemma, this group vanishes. So we can find a map $s : a \mapsto s(a) \in \mathfrak{t}^+_{\mathbf{E}_0}$ such that $t(a,b) = a \cdot s(b) - s(ab) + s(a) \cdot b$. Then $e'_a = e_a + s(a)$ is a lifting of $\bar{\mathbf{E}}$. It remains to prove

LEMMA 1.3. *Let $M$ be an $A$-bimodule. Then* $\mathrm{Ext}^r_{A\otimes A}(A, M) = 0$ *for all* $r \geq 2$.

PROOF. $A \otimes A$ is the coordinate ring of an irreducible affine non singular surface over $\mathbb{F}_q$, see EGA IV, (4.6.5) and (17.10.3). Hence for any closed point, its local ring is regular of dimension 2. We claim that the exact sequence

$$(1) \qquad 0 \to \ker \mu \to A \otimes A \xrightarrow{\mu} A \to 0$$

is a projective resolution of $A$. Let $\mathfrak{Q}$ be a maximal ideal of $A \otimes A$ containing $\ker \mu$, and let $\mathfrak{q} = \mu(\mathfrak{Q})$. Then, over $(A \otimes A)_{\mathfrak{Q}}$, $A_{\mathfrak{q}} = A_{\mathfrak{Q}}$ has cohomological dimension 1. In fact, any element in $\mathfrak{Q} - \ker \mu$ is an $A_{\mathfrak{q}}$-sequence, see [**Se**], IV, 4. Now the formula

$$\mathrm{dh}\, A_{\mathfrak{q}} + \mathrm{codh}\, A_{\mathfrak{q}} = 2 \ ,$$

loc. cit. IV, prop. 21, implies that $(\ker \mu)_{\mathfrak{Q}}$ is free. The lemma follows from this. □

(3) Finally, we compute the tangent space $t_D = D(\ell[\varepsilon])$ of our functor $D$. Any deformation over $\ell[\varepsilon]$ is given by $a \mapsto e_a^{(0)} + \varepsilon \cdot \delta(a)$, where $\delta$ is a derivation from $A$ into the bimodule $\mathfrak{t}^+_{\mathbf{E}_0}$. Two deformations are isomorphic if and only if the two derivations differ by a commutator, i.e., by a map $a \mapsto [v, a]$ for some $v \in \mathfrak{t}_{\mathbf{E}_0}$. Thus $t_D$ is the cokernel of $\mathfrak{t}_{\mathbf{E}_0} \to \mathrm{Der}(A, \mathfrak{t}^+_{\mathbf{E}_0})$, $v \mapsto [v, \cdot]$. Since $\ker \mu$ is the bimodule of the universal derivation and $\mathrm{Hom}_{A\otimes A}(A, \mathfrak{t}_{\mathbf{E}_0}) = 0$, we get from (1) the commutative diagram with exact lines

$$\begin{array}{ccccccc}
0 \to \mathrm{Hom}_{A\otimes A}(A \otimes A, \mathfrak{t}^+_{\mathbf{E}_0}) & \to & \mathrm{Der}_{\mathbb{F}_q}(A, \mathfrak{t}^+_{\mathbf{E}_0}) & \to & \mathrm{Ext}^1_{A\otimes A}(A, \mathfrak{t}^+_{\mathbf{E}_C}) \to 0 \\
\downarrow & & \downarrow & & \downarrow \nu_* \\
0 \to \mathrm{Hom}_{A\otimes A}(A \otimes A, \mathfrak{t}_{\mathbf{E}_0}) & \to & \mathrm{Der}_{\mathbb{F}_q}(A, \mathfrak{t}_{\mathbf{E}_0}) & \to & \mathrm{Ext}^1_{A\otimes A}(A, \mathfrak{t}_{\mathbf{E}_C}) \to 0
\end{array}$$

We see that $t_D = \mathrm{im}(\nu_*)$. But $\nu_*$ sits in the exact sequence

$$(2) \qquad 0 \to \ell \to \mathrm{Ext}^1_{A\otimes A}(A, \mathfrak{t}^+_{\mathbf{E}_0}) \xrightarrow{\nu_*} \mathrm{Ext}^1_{A\otimes A}(A, \mathfrak{t}_{\mathbf{E}_0}) \to \ell \to 0 \, ,$$

so $\dim t_D = \dim \mathrm{Ext}^1_{A\otimes A}(A, \mathfrak{t}_{\mathbf{E}_0}) - 1$.

Since the functor $\mathrm{Ext}^1_{A\otimes A}(A, \cdot)$ is right exact and commutes with direct sums, there is a canonical isomorphism

$$\mathrm{Ext}^1_{A\otimes A}(A, M) \cong M \otimes_{A\otimes A} \mathrm{Ext}^1_{A\otimes A}(A, A \otimes A) \ ,$$

(cf. [**EGA III**], (7.2.5.)). Let denote $\omega$ the second factor on the right. We claim that it is an invertible $A$-module. Since it is annihilated by

$\ker \mu$, it is an $A$-module. But for any prime $\mathfrak{q}$ as above, $\omega_{\mathfrak{q}}$ is computed by dualizing the exact sequence considered in the lemma. This makes the result obvious.

Now we have to compute $\mathrm{Ext}^1_{A\otimes A}(A, \mathfrak{t}_{\mathbf{E}_0}) = \mathfrak{t}_{\mathbf{E}_0} \otimes \omega$. Since $(\ell \otimes A) \otimes_{A\otimes A} A \cong \ell$ and $\omega$ is invertible, we obtain

$$\mathrm{Ext}^1_{A\otimes A}(A, \mathfrak{t}_{\mathbf{E}_0}) \cong \mathfrak{t}_{\mathbf{E}_0} \otimes_{\ell\otimes A} \ell \,.$$

This is $\ell^d$ by the following lemma. Hence $\dim t_D = d - 1$. By smoothness and [**S**], (2.5), the proposition is proved. □

LEMMA 1.4. *The $\ell \otimes A$-module $\mathfrak{t}_{\mathbf{E}_0}$ is projective of rank $d$.*

PROOF. The ring $\ell \otimes A$ is a Dedekind domain, [**EGA IV**], (4.6.3), and the module $\mathfrak{t}_{\mathbf{E}_0}$ is torsion free. In fact a cyclic torsion submodule would have finite dimension over $\ell$, so the degree of its elements would be bounded. This is not possible. Now fix any $a \in A$ with $|a|_\infty > 1$, and let $N = d \cdot \log_q |a|_\infty$. Division by $e_a$ shows that $\mathfrak{t}_{\mathbf{E}_0}$ is generated by all $\sum_{i<N} a_i \tau^i$. Hence it is projective. Moreover, as a $\ell$-vector space, $\mathfrak{t}_{\mathbf{E}_0}/a \cdot \mathfrak{t}_{\mathbf{E}_0}$ has dimension $N$. Thus it has rank $d$ over $\ell \otimes A/a$. □

We give some applications. First we show that quotients by subschemes of division points exist in the category of Drinfeld modules.

PROPOSITION 1.5. *Let $\mathbf{E}$ be a Drinfeld module over a scheme $S$, and let $(0) \neq I \subseteq A$ be an ideal. Then the quotient $\mathbf{E}/\mathbf{E}[I]$ is naturally a Drinfeld module over $S$.*

PROOF. By 1, 3.2, there is a homomorphism $\mathbf{E} \xrightarrow{p_I} \mathbb{G}_{a,B}$ having kernel $\mathbf{E}[I]$, and there is an induced homomorphism $\bar{e} : A \to \mathrm{End}(\mathbb{G}_{a,B})$. The problem is to show that $\partial \circ \bar{e}$ is the structure morphism. It will be sufficient to prove this for $S = \mathrm{Spec}\, B$, where $B$ is the localization of a finitely generated $A$-algebra at a maximal ideal. This can be checked modulo powers of the maximal ideal of $B$, so we are reduced to the case of $B$ being a local artinian ring. Let $\ell = B/\mathfrak{m}_B$, and let $\mathfrak{p} \subseteq A$ be the preimage of $\mathfrak{m}_B$. Then $\mathfrak{p} \neq (0)$. Choose a regular parameter $T$ of $\mathfrak{p}A_{\mathfrak{p}}$. It defines a unique isomorphism $\hat{A}_{\mathfrak{p}} \cong A/\mathfrak{p}[[T]]$, so $O := \ell[[T]]$ is an algebra over $A$. Since $\ell$ is finite over $\mathbb{F}_q$, there is a unique lift of $\ell$ making $\ell$ a subfield of $B$. Thus $B$ is an $O$-algebra. Let $\tilde{\mathbf{E}}$ be the universal deformation of $\mathbf{E}_0 = \mathbf{E} \otimes_B \ell$. It is enough to show that $\tilde{\mathbf{E}}/\tilde{\mathbf{E}}[I]$ is a Drinfeld module. This follows from 2, 3.2, (i) by passing to $Q(O[[T_1, \ldots, T_{d-1}]])$. □

We denote the quotient $\mathbf{E}/\mathbf{E}[I]$ by $p_I : \mathbf{E} \to \mathbf{E}/I$ or $\mathbf{E}/I$ for short.

REMARK 1.6. The proof shows that any local artinian $A$-algebra $B$ of finite type can be considered as as an object of $\mathcal{C}_O$ for some complete discrete valuation ring $O$. Therefore, given a Drinfeld module over $B/\mathfrak{m}_B$, we can speak about its deformations over $B$. These do not depend on the actual choice of an $O$-algebra structure. Moreover, one can assume that $\mathfrak{p} \cdot O = \mathfrak{m}_O$.

We can give a characterization of the functor represented by a quotient Drinfeld module. This result will be important in the next chapter. We begin with a lemma.

LEMMA 1.7. *Let* $\mathbf{E} \xrightarrow{p_I} \mathbf{F} \xrightarrow{q_J} \mathbf{G}$ *be isogenies of Drinfeld modules such that* $\ker p_I = \mathbf{E}[I]$ *and* $\ker q_J = \mathbf{F}[J]$. *Then* $\ker q_J \circ p_I = \mathbf{E}[I \cdot J]$.

PROOF. For any $a \in I$ and $b \in J$ we can draw the following commutative diagram

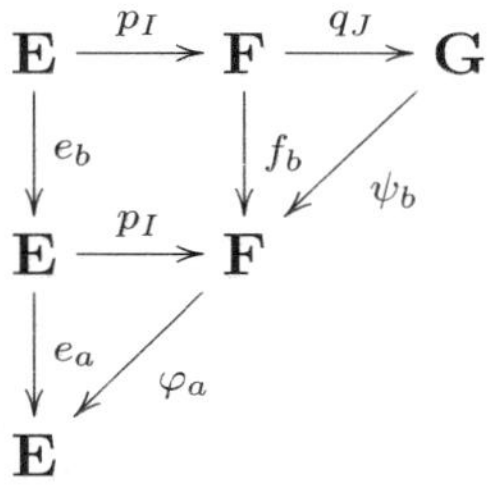


This shows that $\ker e_c \supseteq \ker q_J \circ p_I$ for all $c \in I \cdot J$. Then also $\mathbf{E}[I \cdot J] \supseteq \ker q_J \circ p_I$. Comparing degrees, we must have equality, c.f. 2, 4.1. □

PROPOSITION 1.8. *Let $T$ be a scheme over $S$. Then there is an isomorphism of $A$-modules*

$$(\mathbf{E}/I)(T) \tilde{\rightarrow} \mathrm{Hom}_A(I, \mathbf{E}(T)) .$$

PROOF. Let $\mathbf{E} \xrightarrow{p_I} \mathbf{E}/I$ be the projection, and let $\sigma \in (\mathbf{E}/I)(T)$. Then we associate to it the map $\nu_\sigma(a) := \varphi_a \circ \sigma$, where $\varphi_a$ is as in lemma 1.7. The map $\sigma \mapsto \nu_\sigma$ is injective. In fact, $\varphi_a \circ \sigma = 0$ for all $a \in I$ implies $e_a \circ (\sigma \times_{\mathbf{E}/I} \mathbf{E}) = 0$ for all $a \in I$. Hence $\sigma \times \mathbf{E}$ is a map into $\mathbf{E}[I]$, so $\sigma = 0$.

Conversely, let $I \xrightarrow{\nu} \mathbf{E}(T)$ be a homomorphism. Fix $0 \neq c \in I$, and let $(c) = I \cdot J$. Then $q_J := \varphi_c$ identifies $\mathbf{E}$ with $(\mathbf{E}/I)/J$ by the lemma. Let $c = \sum_i a_i b_i$, with $a_i \in I$ and $b_i \in J$. Put

$$\sigma := \sum_i \psi_{b_i} \circ \nu(a_i) .$$

Since $q_J \circ p_I = e_c$, we get from the diagram in the lemma

$$\begin{aligned}\nu_\sigma(a) &= \sum_i \varphi_a \circ \psi_{b_i} \circ \nu(a_i)\\ &= \sum_i e_{\frac{ab_i}{c}} \circ \nu(a_i)\end{aligned}$$

(As $ab_i \in IJ = (c)$, it follows, that $\frac{ab_i}{c} \in A$)

$$\begin{aligned}&= \sum_i \nu\left(\frac{ab_i a_i}{c}\right)\\ &= \nu(a)\,.\end{aligned}$$

This proves surjectivity. □

COROLLARY 1.9. *For any $A$-module $\Lambda$ there is a canonical isomorphism*

$$\mathrm{Hom}_A(\Lambda, (\mathbf{E}/I)(T)) \tilde{\rightarrow} \mathrm{Hom}_A(I \otimes \Lambda, \mathbf{E}(T))\,.$$

PROOF. Apply $\mathrm{Hom}_A(\Lambda, \cdot)$ to the isomorphism of proposition 1.8. □

REMARK 1.10. Let $\nu \in \mathrm{Hom}_A(I \otimes \Lambda, \mathbf{E}(T))$. The corresponding map $\nu' \in \mathrm{Hom}_A(\Lambda, (\mathbf{E}/I)(T))$ is characterized by the property that for any $a \in I$ the diagram

$$\begin{array}{ccccc} I \otimes \Lambda & \xrightarrow{P\otimes id} & \Lambda & \xrightarrow{\Phi_a \otimes id} & I \otimes \Lambda \\ \downarrow \nu & & \downarrow \nu' & & \downarrow \nu \\ \mathbf{E}(T) & \xrightarrow{p_I} & (\mathbf{E}/I)(T) & \xrightarrow{\varphi_a} & \mathbf{E}(T) \end{array}$$

commutes. Here $P$ denotes the embedding, and $\Phi_a$, $\varphi_a$ are the maps satisfying $\Phi_a \circ P(b) = ab$ and $\varphi_a \circ p_I = e_a$. As above, we have the formula

$$\nu'(\lambda) = \sum_i \psi_{b_i} \circ \nu(a_i \otimes \lambda)\,.$$

As a second application, we will give a different proof of the de Rham isomorphism theorem of [**Ge**]. Let $\mathbb{C}_\infty$ be the completion of an algebraic closure of $k_\infty$, let $\mathbf{E}_0$ be a Drinfeld module over $\mathbb{C}_\infty$ of rank $d$, and let $\Lambda_{\mathbf{E}_0} \subseteq \mathbb{C}_\infty$ be the corresponding lattice, see 2, sec. 1. Recall that the first *Betti cohomology* group of $\mathbf{E}_0$ is defined as

$$H^1(\mathbf{E}_0, \mathbb{C}_\infty) = \mathrm{Hom}_A(\Lambda_{\mathbf{E}_0}, \mathbb{C}_\infty)\,.$$

On the other hand, the first *de Rham cohomology* of $\mathbf{E}_0$ is

$$H^1_{\mathrm{DR}}(\mathbf{E}_0, \mathbb{C}_\infty) = \mathrm{Der}_{\mathbb{F}_q}(A, \mathfrak{t}^+_{\mathbf{E}_0})/\mathrm{Der}(A, \mathfrak{t}^+_{\mathbf{E}_0})_{si}\,,$$

where the quotient is formed with respect to the *strictly inner* derivations, i.e., those of the form $[a, \cdot]$ with $a \in \mathfrak{t}^+_{\mathbf{E}_0}$. Hence by the above discussion, $H^1_{\mathrm{DR}}(\mathbf{E}_0, \mathbb{C}_\infty)$ identifies canonically with $\mathrm{Ext}^1_{A\otimes A}(A, \mathfrak{t}^+_{\mathbf{E}_0})$, so its dimension is $d$. Let $\tilde{\mathfrak{t}}^+_{add}$ be the $A$-bimodule of all $\mathbb{F}_q$-linear series in $X^q\mathbb{C}_\infty[[X]]$ with infinite radius of convergence. Here $a \in A$ acts as $aX$ from both sides. Of course here $X^q\mathbb{C}_\infty[[X]]$ is just another way of writing $\mathbb{C}_\infty[[\tau]] \cdot \tau$ ($= \tau\mathbb{C}_\infty[[\tau]]$ because $\mathbb{C}_\infty$ is perfect) inside the skew power series ring $\mathbb{C}_\infty[[\tau]]$, neglecting the convergence conditions for the moment.

LEMMA 1.11. (i) *Let $e \in \mathbb{C}_\infty[X]$ be an $\mathbb{F}_q$-linear polynomial, and let $v \in \tilde{\mathfrak{t}}^+_{add}$. Then there exists a unique series $w \in \tilde{\mathfrak{t}}^+_{add}$ and a $\mathbb{F}_q$-linear polynomial $r$ of degree less than* $\deg e$ *such that*

$$v = we + r\,.$$

(ii) *Let $v_{|\Lambda_{\mathbf{E}_0}} = 0$. Then there exists $w \in \tilde{\mathfrak{t}}^+_{add}$, such that $v = wu$, where $u$ is the series with simple roots $\Lambda_{\mathbf{E}_0}$, see 2, sec. 1.*

PROOF. Assertion (i) follows from Weierstrass division and a scaling argument, c.f [**BGR**], 6.1.5.

Similarly, in (ii), $v$ is divisible by $u_{\Lambda'}$ for any finite subgroup $\Lambda' \subseteq \Lambda$, see 2, 1.2. The quotients converge to a series $w$ on any disk. □

PROPOSITION 1.12. (Gekeler [Ge]) *There exists a canonical isomorphism*

$$H^1_{\mathrm{DR}}(\mathbf{E}_0, \mathbb{C}_\infty) \to H^1(\mathbf{E}_0, \mathbb{C}_\infty)\,.$$

PROOF. The homomorphism $\mathfrak{t}^+_{\mathbf{E}_0} \to \tilde{\mathfrak{t}}^+_{add}u$ sending $v$ to $v^\sharp(u)$ induces an isomorphism $H^1_{\mathrm{DR}}(\mathbf{E}_0, \mathbb{C}_\infty) = \mathrm{Ext}^1_{A\otimes A}(A, \mathfrak{t}^+_{\mathbf{E}_0})$ to $\mathrm{Ext}^1_{A\otimes A}(A, \tilde{\mathfrak{t}}^+_{add}u)$. This will be proved in a lemma below. On the other hand, by 1.11, (ii), restriction of analytic functions identifies the quotient $\tilde{\mathfrak{t}}^+_{add}/\tilde{\mathfrak{t}}^+_{add}u$ with a subspace $M$ of $\mathrm{Hom}_{\mathbb{F}_q}(\Lambda, \mathbb{C}_\infty)$. Now we have the exact sequence

$$0 \to \mathrm{Hom}_{A\otimes A}(A, M) \to \mathrm{Ext}^1_{A\otimes A}(A, \tilde{\mathfrak{t}}^+_{add}u) \to \mathrm{Ext}^1_{A\otimes A}(A, \tilde{\mathfrak{t}}^+_{add})\,.$$

But $\mathrm{Hom}_{A\otimes A}(A, M)$ is a subspace of $\mathrm{Hom}_{A\otimes A}(A, \mathrm{Hom}_{\mathbb{F}_q}(\Lambda, \mathbb{C}_\infty)) = \mathrm{Hom}_A(\Lambda, \mathbb{C}_\infty)$, and $\mathrm{Ext}^1_{A\otimes A}(A, \tilde{\mathfrak{t}}^+_{add}) = 0$. In fact, let $A \xrightarrow{\delta} \tilde{\mathfrak{t}}^+_{add}$ be a derivation. Fix some non-constant element $a \in A$, and write $\delta(a) = \sum_{i>0} a_i X^{q^i}$. Then $\delta = [v, \cdot]$, with $v = \sum_{i>0} v_i X^{q^i}$ and $v_i = a_i/(a^{q^i} - a)$. This proves vanishing of the Ext-group. Comparing dimensions, we see that $\mathrm{Ext}^1_{A\otimes A}(A, \tilde{\mathfrak{t}}^+_{add}u)$ is isomorphic to $\mathrm{Hom}_A(\Lambda, \mathbb{C}_\infty) = H^1(\mathbf{E}, \mathbb{C}_\infty)$, so the proposition is proved. □

It remains to show

LEMMA 1.13. *The embedding* $\mathfrak{t}^+_{\mathbf{E}_0} \to \tilde{\mathfrak{t}}^+_{add}u$ *induces an isomorphism*

$$\mathrm{Ext}^1_{A\otimes A}(A, \mathfrak{t}^+_{\mathbf{E}_0}) \cong \mathrm{Ext}^1_{A\otimes A}(A, \tilde{\mathfrak{t}}^+_{add}u)\,.$$

PROOF. Consider the quotient $Q = \tilde{\mathfrak{t}}^+_{add}u/\mathfrak{t}^+_{\mathbf{E}_0}$. We claim that $\mathrm{Hom}_{A\otimes A}(A, Q) = 0$. Let $a\bar{v} - \bar{v}a = 0$ for some non-constant $a \in A$. This means that $av - ve^\sharp_a$ is a polynomial in $\mathfrak{t}^+_{\mathbf{E}_0}$. Since $\deg e^\sharp_a > 1$, $v$ must be a polynomial by the general convergence lemma, 1, 4.4. Hence $\bar{v} = 0$. Using this result, we obtain an exact sequence

$$0 \to \mathrm{Ext}^1_{A\otimes A}(A, \mathfrak{t}^+_{\mathbf{E}_0}) \xrightarrow{i} \mathrm{Ext}^1_{A\otimes A}(A, \tilde{\mathfrak{t}}^+_{add}u) \to \mathrm{Ext}^1_{A\otimes A}(A, Q) \to 0\,.$$

But we have seen that $\mathrm{Ext}^1_{A\otimes A}(A, \tilde{\mathfrak{t}}^+_{add}u)$ has dimension $d$, thus $i$ is an isomorphism. □

Of course, this lemma is just a variant of the "injective"-part in Gekeler's original proof.

## 2. Deformations of isogenies

In the same manner we now investigate deformations of isogenies. Let $O$ be a complete noetherian local ring with residue field $\ell$, let $\mathbf{E}_0$ and $\mathbf{F}_0$ be Drinfeld modules of rank $d > 0$ over $\ell$, let $\varphi_0 : \mathbf{E}_0 \to \mathbf{F}_0$ be an isogeny, and let $B \in \mathcal{C}_O$. We identify $\varphi_0$ with a $q$-linear endomorphism of $\mathbb{G}_{a,\ell}$.

DEFINITION 2.1. A deformation of $\varphi_0$ over $B$ is an isogeny $\mathbf{E} \xrightarrow{\varphi} \mathbf{F}$, where $\mathbf{E}$ and $\mathbf{F}$ are deformations of $\mathbf{E}_0$ and $\mathbf{F}_0$ respectively, such that $\ell \otimes_B \varphi = \varphi_0$.

Two deformations are *isomorphic* if there is a commutative diagram

$$\begin{array}{ccc} \mathbf{E} & \xrightarrow{\varphi} & \mathbf{F} \\ \downarrow{\scriptstyle u} & & \downarrow{\scriptstyle v} \\ \mathbf{E}' & \xrightarrow{\varphi'} & \mathbf{F}' \end{array}$$

such that $\ell \otimes_B u$ and $\ell \otimes_B v$ are the identity.

Again, the identity is the only automorphism of a deformation. Associating with $B$ the isomorphism classes of deformations of $\varphi_0$ yields a functor on $\mathcal{C}_O$, which we denote by $\mathrm{Def}_{\varphi_0}$. Any deformation is uniquely represented by an isogeny of standard Drinfeld modules c.f. 2, 2.3. So we assume all Drinfeld modules to be standard.

We set $t_{\mathbf{E}_0} := t_{\mathrm{Def}\mathbf{E}_0}$ and $t_{\varphi_0} := t_{\mathrm{Def}\varphi_0}$ for short. The isogeny $\varphi_0$ induces a map $\mathbf{t}(\varphi_0) : \mathbf{t}_{\mathbf{F}_0} \to \mathbf{t}_{\mathbf{E}_0}$ of $\ell \otimes A$-modules sending $\eta$ to $\eta \cdot \varphi_0$ and preserving $\mathbf{t}^+$. By the exact sequence 1.3, (2), there is a map

$$d\varphi_0 : t_{\mathbf{F}_0} \to t_{\mathbf{E}_0}.$$

The dimension of $\ker(d\varphi_0)$ is called the *defect* of $\varphi_0$, denoted by $\delta(\varphi_0)$.

One shows that $\mathrm{Def}_{\varphi_0}$ is homogeneous by the same arguments given for Drinfeld modules.

Now let us consider the problem of lifting. Let

$$0 \to (\varepsilon) \to B \to \bar{B} \to 0$$

be a small extension and let $\bar{\varphi}$ be a deformation of $\varphi_0$ over $\bar{B}$. We lift $\bar{\mathbf{E}}$ and $\bar{\mathbf{F}}$ to deformations $\mathbf{E}$ and $\mathbf{F}$ over $B$, and take any lifting of $\bar{\varphi}$ to an $\mathbb{F}_q$-linear endomorphism $\varphi$ of $\mathbb{G}_{a,B}$. Then $\varphi$ is a lifting if and only if

$$s(a) := f_a\varphi - \varphi e_a$$

vanishes for all $a \in A$. In any case, $s$ is a 1-cocycle with values in $\mathbf{t}^+_{\mathbf{E}_0}$:

$$s(ab) = f_a \circ s(b) + s(a) \circ e_b = a \cdot s(b) + s(a) \cdot b.$$

If its class $[s]$ in $t_{\mathbf{E}_0} \subset \mathrm{Ext}^1_{A\otimes A}(A, \mathbf{t}_{\mathbf{E}_0})$ vanishes, we can change the lifting $\varphi$ to get an isogeny.

On the other hand, we may change the liftings of $\bar{\mathbf{E}}$ and $\bar{\mathbf{F}}$, say $e'_a = e_a + \eta(a)$ and $f'_a = f_a + \vartheta(a)$. Then we obtain the class

$$s'(a) = s(a) + \vartheta(a)\varphi - \partial\varphi_0 \cdot \eta(a),$$

so, in $t_{\mathbf{E}_0}$, this amounts to

$$[s'] = [s] + d\varphi_0([\vartheta]) - \partial\varphi_0 \cdot [\eta].$$

Thus $\bar{\varphi}$ can be lifted if and only if we can make $[s']$ vanish.

If we take $B = \ell[\varepsilon]$, any liftings $\mathbf{E}$, $\mathbf{F}$, and $\varphi = \varphi_0$, the class $[s]$ is $d\varphi_0([\mathbf{E}]) - \partial\varphi_0 \cdot [\mathbf{F}]$. Here $[\mathbf{E}]$ is the class of $\mathbf{E}$ in $t_{\mathbf{E}_0}$. So there exists a lifting to an isogeny from $\mathbf{E}$ to $\mathbf{F}$ if and only if this class vanishes.

Finally, let $\mathbf{E}$ and $\mathbf{F}$ be the trivial deformations over $\ell[\varepsilon]$. Then the only lifting of $\varphi_0$ is the trivial one.

Altogether we have proved the following

PROPOSITION 2.2. (i) $\mathrm{Def}_{\varphi_0}$ *is pro-represented by a complete noetherian $O$-algebra with residue field $\ell$.*

(ii) *There is an exact sequence*

$$0 \to t_{\varphi_0} \to t_{\mathbf{E}_0} \oplus t_{\mathbf{F}_0} \to t_{\mathbf{E}_0} \to \mathrm{coker} \to 0,$$

*where $(x, y) \in t_{\mathbf{E}_0} \oplus t_{\mathbf{F}_0}$ is mapped to $\partial\varphi_0 \cdot x - d\varphi_0(y)$.*

(iii) *Given a small extension $B \to \bar{B}$ and a deformation $\bar{\varphi}$ over $\bar{B}$, the image of class $[s]$ in* coker *is the obstruction for a lift of $\bar{\varphi}$.*

COROLLARY 2.3. (i) *If* $d\varphi_0$ *is bijective, then the functor* $\mathrm{Def}_{\varphi_0}$ *is smooth, and for any deformation* $\mathbf{E}$ *there exists a unique deformation* $\varphi : \mathbf{E} \to \mathbf{F}$ *of* $\varphi_0$.

(ii) *If* $\partial\varphi_0 \neq 0$*, then the functor* $\mathrm{Def}_{\varphi_0}$ *is smooth, and for any deformation* $\mathbf{F}$ *there exists a unique deformation* $\varphi : \mathbf{E} \to \mathbf{F}$.

(iii) *If* $\partial\varphi_0 \neq 0$*, then* $\dim t_\varphi = d - 1$*. If* $\partial\varphi_0 = 0$*, we have*

$$\dim t_\varphi = d - 1 + \delta(\varphi_0)\,.$$

PROOF. This follows by induction over the length of $B$ using the exact sequence in the proposition. □

In case of (i), there is a natural transformation

$$D(\varphi_0)_* : \mathrm{Def}_{\mathbf{E}_0} \to \mathrm{Def}_{\mathbf{F}_0}\ ,$$

whereas in case of (ii), there is a natural transformation

$$D(\varphi_0)^* : \mathrm{Def}_{\mathbf{F}_0} \to \mathrm{Def}_{\mathbf{E}_0}\ .$$

EXAMPLE 2.4. Let $\mathfrak{p}$ be the characteristic of $\ell$, and let $\varphi_0 = e_a^{(0)}$ for some $a \notin \mathfrak{p}$. Then $d\varphi_0$ is just the multiplication by $a$, thus it is bijective. The same holds for any divisor of $e_a^{(0)}$ in $\mathrm{End}_{\mathbb{F}_q}(\mathbb{G}_{a,\ell})$, which is an isogeny to some Drinfeld module, since generally

$$d(\psi_0 \circ \varphi_0) = d\varphi_0 \circ d\psi_0$$

for composable isogenies. In particular, let $(a) = J \cdot \mathfrak{p}^n$, $(J, \mathfrak{p}) = (1)$, be any principal ideal, and let $e_a^{(0)} = \varphi_0 \circ p_{\mathfrak{p}^n}$, where $p_{\mathfrak{p}^n}$ is the projection onto $\bar{\mathbf{E}}_0 := \mathbf{E}_0/\mathfrak{p}^n$. Then $d\varphi_0$ is bijective. In fact, there exists an isogeny $\psi_0$ such that $\psi_0 \circ \varphi_0 = \bar{e}_b^{(0)}$ and $b \notin \mathfrak{p}$.

## 3. Deformations of level structures

In this section we assume in addition that the morphism $A \xrightarrow{i} O$ is injective. Let $\mathfrak{p} = \mathfrak{m}_O \cap A$. Then we assume that either $\mathfrak{p} = (0)$ or $\mathfrak{p} \cdot O = \mathfrak{m}_O$, i.e. $O$ is unramified over $A_\mathfrak{p}$. Let $(0) \neq I \subseteq A$ be an ideal, let $(\mathbf{E}_0,\, \iota_0)$ be a Drinfeld module of level $I$, and let $R \in \mathcal{C}_O$. A *deformation* of $(\mathbf{E}_0,\, \iota_0)$ is a Drinfeld module of level $I$ over $R$, which specializes to the given one. The corresponding functor is denoted by $\mathrm{Def}_{(\mathbf{E}_0, \iota_0)}$. By definition, isomorphisms of deformations specialize to the identity.

PROPOSITION 3.1. *Let* $(\mathbf{E}_0,\, \iota_0)$ *be a Drinfeld module of rank* $d > 0$ *and of level* $I$*. Then the functor* $\mathrm{Def}_{(\mathbf{E}_0,\, \iota_0)}$ *is pro-represented by a regular algebra* $R_I$ *in* $\hat{\mathcal{C}}_O$ *of dimension* $d$*. For* $J \subseteq I$ *the canonical homomorphism* $R_I \to R_J$ *is finite and flat.*

PROOF. We proceed in several steps.

*Step 1.* The case $\mathfrak{p} \neq 0$ and $I = \mathfrak{p}$. This is the crucial case. Let $\mathbf{E}$ be the universal deformation of $\mathbf{E}_0$, let $\mathbf{E} \xrightarrow{p_\mathfrak{p}} \mathbf{E}/\mathbf{E}[\mathfrak{p}]$ be the projection, let $L$ be a splitting field of $p_\mathfrak{p}^\sharp$ over $Q(R_0)$, and let $R_1 \subseteq L$ be the $R_0$-algebra generated by all roots of $p_\mathfrak{p}^\sharp$. These roots form an $A/\mathfrak{p}$-module of rank $d$, so there exists a lifting of $\iota_0$

$$\begin{array}{ccc} & & \mathbf{E}(R_1) \\ & \overset{\iota}{\nearrow} & \downarrow \\ (\mathfrak{p}^{-1}/A)^d & \xrightarrow{\iota_0} & \mathbf{E}_0(\ell) \end{array}$$

such that $\iota$ is injective. So it is a level structure on $\mathbf{E}_1 := \mathbf{E} \otimes_{R_0} R_1$. We claim that $R_1$ is a regular local ring. Since $R_1$ is a finite domain over the complete ring $R_0$, it is local. It has residue field $\ell$, because $p_\mathfrak{p}^\sharp \bmod \mathfrak{m}_{R_0}$ splits into linear factors. Let $x_1, \ldots, x_d$ be a base of $(\mathfrak{p}^{-1}/A)^d$ such that $x_1, \ldots, x_h$ is a base of $\ker \iota_0$ for some $1 \leq h \leq d$. Let $y_i \in O$ be a lift of $\iota_0(x_i)$ with $y_i = 0$ if $i \leq h$, and let $z_i = \iota(x_i) - y_i$. Consider the ideal $\mathfrak{a} = (z_1, \ldots, z_d) + \mathfrak{m}_{R_1}^2$, and set $\bar{R} = R_1/\mathfrak{a}$. Let $t \in A$ be a regular parameter of $\mathfrak{p}A_\mathfrak{p}$. Since $p_\mathfrak{p}^\sharp$ divides $e_t^\sharp$ and $\mathfrak{p}O = \mathfrak{m}_O$, we have $\mathfrak{m}_O R_1 \subseteq z_1 R_1$, so $\bar{R}$ is isomorphic to $\ell[V]$ for some $\ell$-vector space $V$. Here $\ell[V]$ denotes the algebra $\ell \oplus V$ with $V \cdot V = 0$. Moreover, the map $R_0 \xrightarrow{\pi} \bar{R}$ is surjective. On the other hand, $\mathbf{E} \otimes_{R_0} \bar{R} = \mathbf{E}_1 \otimes_{R_1} \bar{R}$ has an induced level $\mathfrak{p}$ structure for which $x_1, \ldots, x_h$ are mapped to 0. By assertion (iv) in lemma 3.2. below, $\mathbf{E}_0 \otimes_{R_0} \bar{R}$ is trivial. Since $\pi$ is surjective, we conclude that $V = 0$, hence $R_1$ is regular. It is flat over $R_0$ invoking IV, prop. 22, cor. in [**Se**].

Let $(\mathbf{E}', \iota')$ be a deformation over $B \in \mathcal{C}_O$. We choose a coefficient field $\ell \to O$ making any $O$-algebra into a $\ell$-algebra. Then $R_1 \cong \ell[[z_1, \ldots, z_d]]$, and the homomorphism $\rho$ of $\ell$-algebras sending $z_i$ to $\iota'(x_i) - y_i$ yields a Drinfeld module $\mathbf{E}''$ with respect to a possibly different $A$-algebra structure on $B$. But $\mathbf{E}'[\mathfrak{p}] = \mathbf{E}''[\mathfrak{p}] \subseteq \mathbb{G}_{a,B}$. According to assertion (iii) in the lemma below, this implies $\mathbf{E}' = \mathbf{E}''$. In particular, $\rho$ is $A$-linear. Since $O$ is unramified over $A_\mathfrak{p}$, $\ell \cdot A$ is dense in $O$, so $\rho$ is $O$-linear. Hence $(\mathbf{E}_1, \iota)$ is the universal deformation of level $\mathfrak{p}$. We are left to prove:

LEMMA 3.2. *Let $\ell$ be a field over $A$ with characteristic $\mathfrak{p} \neq (0)$, let $V$ be a vector space over $\ell$, let $B$ be a local artinian ring with residue field $\ell$, and let $\mathbf{E}_0$ be a Drinfeld module over $\ell$.*

(i) *Let $\mathbf{E}$ be a deformation of $\mathbf{E}_0$ over $\ell[V]$. If $\mathbf{E}[\mathfrak{p}] = \mathbf{E}_0[\mathfrak{p}] \otimes \ell[V] \subseteq \mathbb{G}_{a,\ell[V]}$, then $\mathbf{E}$ is the trivial deformation.*

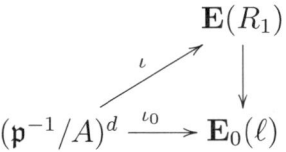

(ii) *Let* $\mathbf{E}$ *and* $\mathbf{E}'$ *be deformations over* $B$. *If* $\mathbf{E}[\mathfrak{p}] = \mathbf{E}'[\mathfrak{p}]$, *then* $\mathbf{E} = \mathbf{E}'$.

(iii) *Let* $A \overset{e^{(1)}}{\underset{e^{(2)}}{\rightrightarrows}} \mathrm{End}(\mathbb{G}_{a,B})$ *be homomorphisms defining Drinfeld modules* $\mathbf{E}_1$ *and* $\mathbf{E}_2$ *with respect to* $\partial \circ e^{(1)}$ *and* $\partial \circ e^{(2)}$, *respectively. Assume that* $\mathbf{E}_1 \otimes_B \ell = \mathbf{E}_2 \otimes_B \ell$ *and* $\mathbf{E}_1[\mathfrak{p}] = \mathbf{E}_2[\mathfrak{p}]$ *as subgroups of* $\mathbb{G}_{a,B}$. *Then* $e^{(1)} = e^{(2)}$.

(iv) *Let* $(\mathbf{E}_0, \iota_0)$ *be a Drinfeld module of level* $\mathfrak{p}$, *and let* $(\mathbf{E}, \iota)$ *be a deformation over* $\ell[V]$, *such that* $\iota_{|\ker \iota_0} = 0$. *Then* $\mathbf{E}$ *is the trivial deformation.*

PROOF. Let $\mathbf{E}_0 \overset{p_\mathfrak{p}^{(0)}}{\longrightarrow} \bar{\mathbf{E}}_0 = \mathbf{E}_0/\mathbf{E}_0[\mathfrak{p}]$ be the projection. Then $\mathbf{E}_0 \otimes \ell[V] \overset{p_\mathfrak{p}^{(0)}\otimes 1}{\longrightarrow} \bar{\mathbf{E}}_0 \otimes \ell[V]$ can be considered as the projection $\mathbf{E} \to \bar{\mathbf{E}} = \mathbf{E}/\mathbf{E}[\mathfrak{p}]$. Let $e_a = e_a^{(0)} \otimes 1 + \delta(a)$, with $\delta(a) \in V$. Then the equations

$$\bar{e}_a \circ (p_\mathfrak{p}^{(0)} \otimes 1) = p_\mathfrak{p}^{(0)} \otimes 1 \circ (e_a^{(0)} \otimes 1 + \delta(a)) = (p_\mathfrak{p}^{(0)} \circ e_a^{(0)}) \otimes 1$$

imply that $\bar{e}_a = \bar{e}_a^{(0)} \otimes 1$, so $\bar{\mathbf{E}}$ is the trivial deformation. On the other hand, there exists an isogeny $\bar{\mathbf{E}} \overset{\varphi}{\to} \mathbf{E}$, such that $\varphi \circ p_\mathfrak{p}^{(0)} \otimes 1 = e_c$ for some $c \in \mathfrak{p} - \mathfrak{p}^2$. By 2.4, $d\varphi$ is bijective. Applying 2.3, (ii), we conclude that $\mathbf{E}$ is trivial. This proves (i).

Assertion (ii) follows by induction over the length of $B$. In fact, let $(\varepsilon) \to B \to \bar{B}$ be a small extension, and assume that $\mathbf{E} \otimes \bar{B} = \mathbf{E}' \otimes \bar{B}$. Then the difference of $\mathbf{E}$ and $\mathbf{E}'$ is a deformation over $\ell[\varepsilon]$ with the same scheme of $\mathfrak{p}$-division points as the trivial one. Now we apply (i).

For (iii), we can again assume that $\mathbf{E}_1 \otimes \bar{B} = \mathbf{E}_2 \otimes \bar{B}$. The quotient $\mathbb{G}_{a,B} \overset{p_\mathfrak{p}}{\to} \mathbb{G}_{a,B}/\mathbf{E}_1[\mathfrak{p}]$ can be considered as a quotient of both $\mathbf{E}_1$ and $\mathbf{E}_2$. Since $p_\mathfrak{p} \otimes \ell$ has height greater than 1, we have $0 = p_\mathfrak{p} \circ (e_a^{(1)} - e_a^{(2)}) = (\bar{e}_a^{(1)} - \bar{e}_a^{(2)}) \circ p_\mathfrak{p}$, $a \in A$. Hence $\bar{e}^{(1)} = \bar{e}^{(2)}$. It follows that $\partial \circ e^{(1)} = \partial \circ e^{(2)}$, so $\mathbf{E}_1$ and $\mathbf{E}_2$ are deformations over the same $A$-algebra. These are the same by (ii).

For (iv), consider again the projection $\mathbf{E} \overset{p_\mathfrak{p}}{\to} \mathbf{E}/\mathbf{E}[\mathfrak{p}]$. Write $p_\mathfrak{p} = p_\mathfrak{p}^{(0)} \otimes 1 + \delta$, and let $p_\mathfrak{p}^\sharp = p_\mathfrak{p}^{(0)\sharp} \otimes 1 + \delta^\sharp$ be the corresponding polynomial. Since $p_\mathfrak{p}^{(0)}$ has height greater than 1, each root of $p_\mathfrak{p}^{(0)\sharp}$ is also a root of $\delta^\sharp$. Moreover, by hypothesis, the multiplicities coincide. Note that all roots of an additive polynomial have the same multiplicity. Hence $p_\mathfrak{p}^\sharp = (1 + v) p_\mathfrak{p}^{(0)\sharp} \otimes 1$ for some $v \in V$. By (i), $\mathbf{E}$ is trivial. The lemma is proved. □

*Step 2.* The case $\mathfrak{p} \neq (0)$ and $I = \mathfrak{p}^n$. Let $(\mathbf{E}_0, \iota_0)$ over $\ell$ be given. Fix $c \in \mathfrak{p} - \mathfrak{p}^2$. Then, for $1 \leq r \leq n$, we construct data

$$R_r,\ (\mathbf{E}_r,\ \iota_r),\ x_1, \ldots, x_d,\ y_1, \ldots, y_d,$$

such that

(a) $R_r$ is a regular algebra in $\hat{\mathcal{C}}_O$;
(b) $(\mathbf{E}_r,\ \iota_r)$ is a universal deformation of level $\mathfrak{p}^r$, and its base ring is $R_r$;
(c) $x_1, \ldots, x_d$ is a base of $(\mathfrak{p}^{-r}/A)^d$, and $x_1, \ldots, x_h$ is a base of $\ker(\iota_0$ restricted to $(\mathfrak{p}^{-r}/A)^d)$;
(d) $y_i$ is a lifting to $O$ of $\iota_0(x_i)$, and $y_i = 0$ for $i \leq h$;
(e) $\iota_r(x_i) - y_i$, $1 \leq i \leq d$, is a regular parameter system in $R_r$;
(f) $\mathfrak{m}_O R_r \subseteq \iota_r(x_1) \cdot R_r$.

For $r = 1$ all data were constructed during the proof of step 1. Let $1 \leq r < n$, and let all data be constructed for $r$. Use 2, 2.4 and 2, 4.3 to see that there is a base $\tilde{x}_1, \ldots, \tilde{x}_d$ of $(\mathfrak{p}^{-r-1}/A)^d$, such that $c \cdot \tilde{x}_i = x_i$ and $\tilde{x}_1, \ldots, \tilde{x}_h$ is a base of $\ker(\iota_0$ restricted). Choose $\tilde{y}_i$ as required in (d). After that put

$$R_{r+1} = R_r[[S_1, \ldots, S_d]]/\mathfrak{a}\ ,$$

where $\mathfrak{a}$ is generated by $e_c^\sharp(S_i) - \iota_r(x_i) + e_c^\sharp(\tilde{y}_i)$, $1 \leq i \leq d$. Let $s_i$ be the class of $S_i$. Since $\iota_r(x_1) \in s_1 \cdot R_{r+1}$, and by hypothesis (f), we have $\mathfrak{m}_O R_{r+1} \subseteq s_1 \cdot R_{r+1}$. Consequently $\iota_r(x_i) - y_i = e_c^\sharp(s_i) + e_c^\sharp(\tilde{y}_i) - y_i$ is in $(s_1, s_i) \cdot R_{r+1}$. Thus $s_1, \ldots, s_d$ is a generating system of $\mathfrak{m}_{R_{r+1}}$, so $R_{r+1}$ is regular (and in $\hat{\mathcal{C}}_O$). Moreover, it is finite over $R_r$ and therefore flat.

We claim that there is a level $\mathfrak{p}^{r+1}$ structure on $\mathbf{E}_{r+1} := \mathbf{E}_r \otimes R_{r+1}$ that maps $\tilde{x}_i$ to $s_i + \tilde{y}_i$. We have to prove that $s_i + \tilde{y}_i$ is annihilated by $\mathfrak{p}^{r+1}$. There exists an element $b \in A - \mathfrak{p}$, such that $b \cdot \mathfrak{p}^{r+1} \subseteq c \cdot \mathfrak{p}^r$, so $s_i + \tilde{y}_i$ is annihilated by $b \cdot \mathfrak{p}^{r+1}$. But $e_b^\sharp$ is étale, so 0 is its only root in $\mathfrak{m}_{R_{r+1}}$. Hence for any $a \in \mathfrak{p}^{r+1}$ , any root of $e_{ab}^\sharp$ lying in the maximal ideal is a root of $e_a^\sharp$. This proves the result.

It remains to show that $(\mathbf{E}_{r+1},\ \iota_{r+1})$ is universal. One argues similarly as in step 1.

*Step 3.* The general case. We write $I = J \cdot \mathfrak{p}^n$ such that $J \not\subseteq \mathfrak{p}$, and where $n = 0$ if $\mathfrak{p} = (0)$. Then $\mathbf{E}_0[J]$ is étale, and $\iota_0$ restricts to an isomorphism $\ell \times (J^{-1}/A)^d \tilde{\rightarrow} \mathbf{E}_0[J]$. Let $(\mathbf{E},\ \iota)$ be a deformation of $(\mathbf{E}_0,\ \iota_0')$ over $B$, where $\iota_0' = \iota_0|(\mathfrak{p}^{-n}/A)^d$. There is a unique lifting of $\iota_0|(J^{-1}/A)^d$ to an isomorphism $B \times (J^{-1}/A)^d \tilde{\rightarrow} \mathbf{E}[J]$. Therefore the restriction morphism

$$\mathrm{Def}_{(\mathbf{E}_0,\, \iota_0)} \rightarrow \mathrm{Def}_{(\mathbf{E}_0,\, \iota_0')}$$

is an isomorphism. Thus the left functor is pro-represented by $R_n = R_{\mathfrak{p}^n}$.

The last statement is clear by construction and [**Se**], loc. cit. The proof of the proposition is complete. □

As an application we will prove that the equality of divisors in the definition of a level $I$ structure holds for all ideals containing $I$.

PROPOSITION 3.3. *Let $(0) \neq I$ be an ideal in $A$, and let $(\mathbf{E}, \iota)$ be a Drinfeld module of rank $d$ and level $I$ over an $A$-scheme $S$. Then we have*

$$\mathbf{E}[I] = \sum_{x \in (I^{-1}/A)^d} \iota(x)\,.$$

PROOF. By similar arguments as in 1.5, it will be sufficient to prove this for the universal Drinfeld module over $R_I$ as discussed above. But in this case, as $R_I$ is a regular algebra in $\hat{\mathcal{C}}_O$, the equality can be checked at the generic point of $\operatorname{Spec} R_I$, where it follows from 2, 4.3, (1). □

## 4. Smoothness of the moduli spaces

Now we are prepared to prove the main result of this chapter. Let $(0) \neq I \subseteq A$ be an ideal such that $|V(I)| \geq 2$, and let $d > 0$.

THEOREM 4.1. (i) *The moduli scheme $M_I^d$ is regular of dimension $d$.*

(ii) *Let $J \subseteq I$. Then the morphism $f : M_J^d \to M_I^d$ given by restriction of level structures is surjective, flat and finite.*

(iii) *The morphism $M_I^d \to \operatorname{Spec} A$ is surjective and flat of relative dimension $d-1$, and it is smooth over* $\operatorname{Spec} A - V(I)$.

(iv) *$M_I^1$ is finite over* $\operatorname{Spec} A$

PROOF. Let $\mathfrak{m}$ be a closed point of $M = M_I^d$, and let $\mathfrak{p} = A \cap \mathfrak{m}$ be its image in $\operatorname{Spec} A$. Since $\mathcal{O}(M)$ is finitely generated, we have $\mathfrak{p} \neq (0)$. By 1.6, there exists a factorization $A_{\mathfrak{p}} \to O \to \hat{\mathcal{O}}_{M,\mathfrak{m}}$, where $O$ is a complete discrete valuation ring with residue field $\ell = \mathcal{O}(M)/\mathfrak{m}$, and such that $\mathfrak{p} \cdot O = \mathfrak{m}_O$. Moreover, since $\ell$ is algebraic over $A/\mathfrak{p}$, $O$ is finite and étale over $\hat{A}_{\mathfrak{p}}$.

Let $(\mathbf{E}_0, \iota_0)$ be the Drinfeld module over $\ell$ corresponding to the point $\mathfrak{m}$. Then $\hat{\mathcal{O}}_{M,\mathfrak{m}}$ pro-represents the functor $\operatorname{Def}_{(\mathbf{E}_0,\iota_0)}$. In fact, any deformation of $(\mathbf{E}_0, \iota_0)$ over some $B \in \mathcal{C}_O$ arises from an $A$-algebra homomorphism $\hat{\mathcal{O}}_{M,\mathfrak{m}} \to B$, which is $O$ linear, since $O$ is étale over $\hat{A}_{\mathfrak{p}}$. By 3.1, $\hat{\mathcal{O}}_{M,\mathfrak{m}}$ is regular of dimension $d$, so the same holds for $\mathcal{O}_{M,\mathfrak{m}}$ as well.

Let $\mathbf{E}_0$ be a Drinfeld module over an algebraically closed field. From 2, 4.3, any level $I$ structure on it can be lifted to a level $J$ structure. This proves surjectivity of $f$.

To check that it is finite, recall the construction of $\mathcal{O}(M_J^d)$, see 2, 5.1. We see that it will be sufficient to show that $\iota(x)$ is integral over $\mathcal{O}(M_I^d)$ for $x \in (J^{-1}/A)^d$. But this follows from $e_a^\sharp(\iota(x)) = 0$ for any $a \in J$.

It remains to verify flatness. This follows by standard techniques from the flatness of $\hat{\mathcal{O}}_{M_I,f(\mathfrak{m})} \to \hat{\mathcal{O}}_{M_J,\mathfrak{m}}$ for any closed point $\mathfrak{m}$ of $M_J^d$ by 3.1.

To prove smoothness outside of $V(I)$, we must show that for any $\mathfrak{p} \notin V(I)$, $\hat{\mathcal{O}}_{M,\mathfrak{m}}$ is smooth over $\hat{A}_\mathfrak{p}$. But $O$ is finite and étale over $\hat{A}_\mathfrak{p}$ and $\hat{\mathcal{O}}_{M,\mathfrak{m}}$ is isomorphic to $O[[T_1, \ldots, T_{d-1}]]$, see step 3 of the proof of 3.1.

To prove flatness at a point $\mathfrak{p} \in \operatorname{Spec} A$, consider the diagram

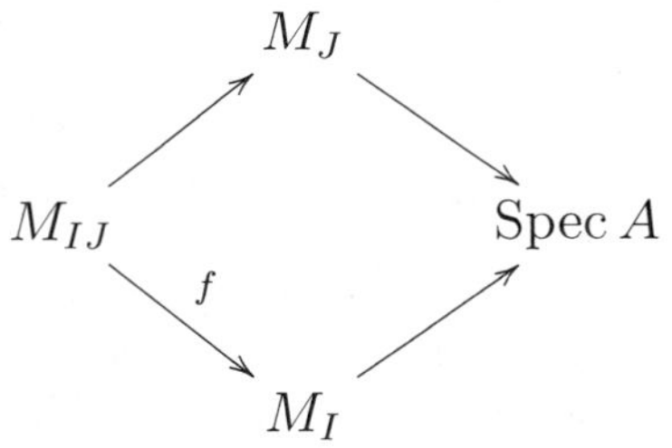


where $V(I) \cup \mathfrak{p}$ and $V(J)$ are disjoint. By smoothness and (ii), the composition is flat over $\operatorname{Spec} A - V(J)$, and $f$ is faithfully flat.

The proof of surjectivity and of (iv) is postponed until 4, 2.3. □

COROLLARY 4.2. *The limit* $M^d = \varprojlim M_I^d$ *is a normal scheme that is faithfully flat over* $\operatorname{Spec} A$.

## 5. Group action on the moduli space

Let $\mathbb{A}_\mathrm{f}$ denote the ring of finite adeles of $k$, i.e.

$$\mathbb{A}_\mathrm{f} = \prod_{\mathfrak{p} \neq \infty}{}' k_\mathfrak{p} ,$$

where the prime means that all but finitely many components lie in $\hat{A}_\mathfrak{p}$. It contains the ring $\hat{A} := \prod \hat{A}_\mathfrak{p}$, where the product is taken over all $\mathfrak{p} \neq \infty$.

Let $|V(I)| \geq 2$ , let $(\mathbf{E}, \iota)$ be an $I$-level Drinfeld module of rank $d$ over $S$ with bundle $\mathcal{L}$, and let $H \subseteq (I^{-1}/A)^d$ be a submodule. Then

$$\operatorname{div}(H) = \sum_{x \in H} \iota(x)$$

is a subgroup of $\mathbb{G}_{a,\mathcal{L}}$ which is finite and flat over $S$. Hence $\mathbb{G}_{a,\mathcal{L}}/H \cong \mathbb{G}_{a,\mathcal{L}'}$ and $e$ induces a homomorphism $e'$ from $A$ to $\operatorname{End}(\mathbb{G}_{a,\mathcal{L}'})$.

PROPOSITION 5.1. *$e'$ defines a Drinfeld module of rank d which we denote by $\mathbf{E}/H$. Moreover let*

$$j : (J^{-1}/A)^d \longrightarrow (I^{-1}/A)^d/H$$

*be an injective homomorphism for some ideal $J \supseteq I$. Then the composite*

$$(J^{-1}/A)^d \longrightarrow (I^{-1}/A)^d/H \longrightarrow \mathbf{E}/H(S)$$

*is a $J$-level structure.*

PROOF. It will be sufficient to prove this for the universal Drinfeld module of level $I$ at generic points of $M_I^d$. In this case the first statement follows from 2, 3.2. Moreover, the composite map is injective, so the desired equality of divisors follows by comparing degrees. □

Let $\alpha$ be a matrix in $\mathrm{GL}_d(\mathbb{A}_\mathrm{f})$ with entries in $\hat{A}$. It defines an endomorphism of $(k/A)^d$ with finite kernel, say $H$. Let $(\mathbf{E}, \iota)$ be a total level Drinfeld module of rank $d$ over $S$. Then $\mathbf{E}/H$ is a total level Drinfeld module by virtue of the commutative diagram

$$\begin{array}{ccc} (k/A)^d & \xrightarrow{\iota} & \mathbf{E}(S) \\ \downarrow{\scriptstyle \alpha} & & \downarrow \\ (k/A)^d & \xrightarrow{\iota'} & \mathbf{E}/H(S) \end{array}$$

and proposition 5.1. This construction, being functorial in $S$, defines an action of the monoid $\mathrm{GL}_d(\mathbb{A}_\mathrm{f}) \cap \mathrm{Mat}_d(\hat{A})$ on $M^d$. In particular, let $a$ be in $A - \{0\}$, and let $\alpha := a \cdot id$. Then the above diagram becomes

$$\begin{array}{ccc} (k/A)^d & \xrightarrow{\iota} & \mathbf{E}(S) \\ \downarrow{\scriptstyle a \cdot id} & & \downarrow \\ (k/A)^d & \xrightarrow{\iota} & \mathbf{E}(S) \end{array} .$$

This shows that $A - \{0\}$ acts trivially on $M^d$. Using the canonical isomorphism

$$(\mathrm{GL}_d(\mathbb{A}_\mathrm{f}) \cap \mathrm{Mat}_d(\hat{A}))/(A - \{0\}) \cong \mathrm{GL}_d(\mathbb{A}_\mathrm{f})/k^\times$$

we get an action of $\mathrm{GL}_d(\mathbb{A}_{\mathrm{f}})/k^\times$ on $M^d$.

In the same way, one defines an action of $\mathrm{GL}_d(\hat{A})/\mathbb{F}_q^\times$ on $M_I^d$, which is compatible with restriction maps.

Let $(0) \neq I \subseteq A$ be an ideal. The set of matrices in $\mathrm{GL}_d(\hat{A})$ which are congruent to 1 mod $I$ is called the *$I$-congruence subgroup*, and is denoted by $\Gamma(I)$. For $J \subseteq I$ we set

$$\Gamma_{I,J} := \Gamma(I)/\Gamma(J).$$

The canonical sequence

$$0 \longrightarrow \Gamma(I) \longrightarrow \mathrm{GL}_d(\hat{A}) \longrightarrow \mathrm{GL}_d(A/I) \longrightarrow 0$$

is exact. In fact the surjectivity of the right hand map can be checked locally, in which case it is obvious.

PROPOSITION 5.2. *There is a canonical isomorphism*

$$M_I^d \cong \Gamma(I)\backslash M^d.$$

PROOF. Since $\Gamma_{I,J}$ acts trivially on $M_I^d$, there is a canonical map

$$\Gamma_{I,J}\backslash M_J^d \longrightarrow M_I^d,$$

which is finite. Both schemes are normal. Since $M_I^d$ is flat over $\operatorname{Spec} A$, the images of all components of $M_I^d$ meet $\operatorname{Spec} A - V(J)$. Therefore it is enough to show that $M_I' := M_I^d \times_{\operatorname{Spec} A} (\operatorname{Spec} A - V(J))$ is isomorphic to $\Gamma_{I,J}\backslash M_J'$.

We claim that $M_J'$ is a principal homogeneous space over $M_I'$ for $\Gamma_{I,J}$. In fact, it is flat over $M_I'$, and for any scheme $S$ over $\operatorname{Spec} A - V(I)$, we can interpret $(M_J' \times_{M_I'} M_J')(S)$ as the set of triples $(\mathbf{E}, \iota, \iota')$ consisting of a Drinfeld module over $S$ endowed with $J$-level structures $\iota, \iota'$, whose restrictions to $(I^{-1}/A)^d$ coincide. But $\iota$ and $\iota'$ are isomorphisms

$$S \times (J^{-1}/A)^d \to \mathbf{E}[J].$$

Hence $\iota^{-1}\iota$ is a section in $S \times \Gamma_{I,J}$. Thus the correspondence

$$((\mathbf{E}, \iota),\ (\mathbf{E}, \iota')) \mapsto (\iota^{-1}\iota'; \mathbf{E}, \iota)$$

defines an isomorphism

$$M_J' \times_{M_I'} M_J' \cong \Gamma_{I,J} \times M_J'.$$

Oviously the above correspondence is $\Gamma_{I,J}$-equivariant, where $\Gamma_{I,J}$ is acting on the first factor. Thus we have an isomorphism of $\Gamma_{I,J}$-space as desired.

In particular, we have

$$\Gamma_I\backslash M_J^d \cong M_I^d,$$

which means that

$$\mathcal{O}(M_I^d) \cong \mathcal{O}(M_J^d)^{\Gamma(I)}.$$

Since taking invariants commutes with direct limits, we are done. □

Since $\mathbb{A}_{\mathrm{f}}^{\times}$ is contained in the normalizer of $\Gamma(I)$ in $\mathrm{GL}_d(\mathbb{A}_{\mathrm{f}})$, it acts on $M_I^d$, also. We denote the corresponding action as $\mathbf{E} \mapsto \mathbf{a}_* \mathbf{E}$ resp. $(\mathbf{E}, \iota) \mapsto \mathbf{a}_*(\mathbf{E}, \iota)$ for $a \in \mathbb{A}_{\mathrm{f}}^{\times}$, $(\mathbf{E}, \iota)$ a level $I$ Drinfeld module. For $\mathbf{a} \in \mathbb{A}_{\mathrm{f}}^{\times} \cap \hat{A}$ this action can be described explicitly as follows:

PROPOSITION 5.3. *Let* $\mathbf{a} \in \mathbb{A}_{\mathrm{f}}^{\times} \cap \hat{A}$, *let* $[\mathbf{a}] = \mathbf{a} \cdot \hat{A} \cap A$, *and let* $(\mathbf{E}, \iota)$ *be a level* $I$ *Drinfeld module over* $S$. *Then, as a Drinfeld module,* $\mathbf{a}_* \mathbf{E}$ *is isomorphic to* $\mathbf{E}/[\mathbf{a}]$. *Let* $[\mathbf{a}]$ *be coprime to* $I$, *then* $\mathbf{a}_*(\mathbf{E}, \iota) = (\mathbf{E}/[\mathbf{a}], p_{[\mathbf{a}]} \circ \iota \circ \mathbf{a}^{-1})$.

PROOF. It will be sufficient to prove the second statement. Let $S \to M_I^d$ be the morphism corresponding to $(\mathbf{E}, \iota)$, and let $S' := S \times_{M_I^d} M^d$. Then $S'$ is faithfully flat over $S$, and for $S'$, the assertion follows directly from the definition of the group action. So it holds for $S$ as well. □

It follows from propositions 5.2., 5.3., that the group $\mathrm{GL}_d^0 := \mathrm{GL}_d(\hat{A}) \cdot \mathbb{A}_{\mathrm{f}}^{\times}$ acts on $M_I^d$. Thereby, $\Gamma(I) \cdot k^{\times}$ is acting trivially.

5.4. If $\alpha \in \mathrm{GL}_d(\hat{A})$, we have

$$\alpha_*(\mathbf{E}, \iota) = (\mathbf{E}, \iota \circ \alpha^{-1}) .$$

Let $(\mathbf{E}^{\mathrm{univ}}, \iota)$ be the universal Drinfeld module. Then the above equality corresponds to an isomorphism $(\mathbf{E}, \iota \circ \alpha^{-1}) \cong \mathbf{E}^{\mathrm{univ}} \times_{\alpha} M_I^d$. This defines an action of $\mathrm{GL}_d(\hat{A})$ on $\mathbf{E}^{\mathrm{univ}}$ by means of isomorphisms over the action on $M_I^d$.

5.5. Finally, we set $M_0^d := \mathrm{GL}_d(\hat{A}) \backslash M_I^d$, for some ideal $I$. It follows from prop. 5.2., that this is independent of the choice of $I$. Outside of $V(I)$, the canonical projection $M_I^d \to M_0^d$ is étale. This implies that $M_0^d$ is smooth over $\mathrm{Spec}\, A$. Moreover, the group $\mathbb{A}_{\mathrm{f}}^{\times}/(\hat{A}^{\times} \cdot k^{\times})$ acts on it. Note that the latter is just the class group of $k$, see 5, 1.9.

CHAPTER 4

# Tate Uniformization

In this chapter we begin the construction of a compactification of $M_I^2$ as it was done in [**Dr**], sec. 9. This means to embed the moduli space into a space, which is proper over Spec $A$. As an intermediate step towards compactification we develop the general concept of a formal boundary. Then we show that the failure of $M_I^d$ being proper can be described by means of Tate data. The construction of a formal boundary and of the compactification will be carried out in the next chapter.

## 1. Formal schemes

In this section we recall some facts on formal schemes. We only need noetherian adic formal schemes. For details and a more general concept the reader is referred to [**EGA I**], chap. 10; see also [**Ha**], II, 9.

Let $R$ be a noetherian ring and let $I \subseteq R$ be a fixed ideal. The *completion* of $R$ with respect to the $I$-adic topology on $R$ is the ring $\hat{R} = \varprojlim R/I^n$. $R$ is *complete*, if the canonical homomorphism $R \to \hat{R}$ is bijective. Let $R = (R, I)$ be complete. Then we define $\operatorname{Spf}(R, I)$ or $\operatorname{Spf} R$ for short, the *formal spectrum* of $R$, as the subspace of open prime ideals in Spec $R$. It is the set of primes containing $I$. For $f \in R$ we define

$$R_{\{f\}} := \hat{R}_f\,,$$

completion with respect the $IR_f$-adic topology. For $f\,, g \in R$, there is a canonical homomorphism $R_{\{f\}} \to R_{\{fg\}}$. Similarly, for an open prime ideal $\mathfrak{p}$ we set $R_{\{\mathfrak{p}\}} := \hat{R}_{\mathfrak{p}}$, completion with respect to $IR_{\mathfrak{p}}$. Let $\mathfrak{D}(f) := D(f) \cap \operatorname{Spf} R$. There is a unique sheaf $\mathcal{O}_{\operatorname{Spf} R}$ on $\operatorname{Spf} R$ such that

$$\mathcal{O}_{\operatorname{Spf} R}(\mathfrak{D}(f)) = R_{\{f\}}\,.$$

With this sheaf, $\operatorname{Spf} R$ is a locally ringed space. It is called an *affine formal scheme*.

DEFINITION 1.1. A *noetherian formal scheme* is a locally ringed space $(\mathfrak{X}, \mathcal{O}_{\mathfrak{X}})$ which has a finite open cover $(\mathfrak{U}_i)$ such that for each $i$, the pair $(\mathfrak{U}_i, \mathcal{O}_{\mathfrak{X}|\mathfrak{U}_i})$ is isomorphic to $(\mathrm{Spf}\,(R_i, I_i), \mathcal{O}_{\mathrm{Spf}\,R_i})$ for some complete noetherian ring $R_i$. A *morphism* of formal schemes is a morphism of locally ringed spaces.

REMARK. The ring of sections of an affine formal scheme is a topological ring. Using open covers, one can show that $\mathcal{O}_{\mathfrak{X}}$ is a sheaf of topological rings in a canonical way, see [**EGA**], loc. cit. Moreover, morphisms induce continuous maps of structure sheaves.

EXAMPLES 1.2. (1) If $I = (0)$ we have $\mathrm{Spf}\,R = \mathrm{Spec}\,R$. More generally, any noetherian scheme can be considered as a formal scheme. A formal scheme that is a scheme is called *discrete.*

(2) Let $X$ be a noetherian scheme, and let $Y$ be a closed subscheme of $X$. Let $\mathcal{I}_Y$ be the ideal sheaf of $Y$. Then the *completion* of $X$ along $Y$ is the formal scheme

$$X_{/Y} := (Y, \varprojlim \mathcal{O}_X/\mathcal{I}_X^n)\,.$$

There is a canonical morphism $X_{/Y} \to X$. The completion actually depends only on the closed subset $Y$, and not on the particular scheme structure of $Y$.

(3) In particular, let $O$ be a complete discrete valuation ring. Then there is a canonical morphism of formal schemes

$$\mathrm{Spf}\,O = \mathrm{Spf}\,(O, \mathfrak{m}_O) \to \mathrm{Spec}\,O = \mathrm{Spf}\,(O, (0))\,.$$

Let $\mathfrak{X} \xrightarrow{f} \mathfrak{S}$ be a morphism of noetherian formal schemes. It is called of *finite type* if there exists an affine cover $(\mathfrak{S}_i) = \mathrm{Spf}\,(R_i, I_i)$ and a finite affine cover $(\mathfrak{U}_{ij})$ of $f^{-1}(\mathfrak{S}_i)$ such that $\mathcal{O}(\mathfrak{U}_{ij})$ is a quotient of a ring $R_i\widehat{[T_1, \ldots, T_n]}$, where the polynomial ring is completed with respect to the $I_iR_i[T_1, \ldots, T_n]$-adic topology. In this case, $\mathfrak{X}$ is called of finite type over $\mathfrak{S}$.

(4) Let $\mathfrak{X} \xrightarrow{f} \mathfrak{S}$ and $\mathfrak{Y} \xrightarrow{g} \mathfrak{S}$ be two noetherian formal schemes over the noetherian formal scheme $\mathfrak{S}$. Assume that $\mathfrak{Y}$ is of finite type over $\mathfrak{S}$. Then the fiber product $\mathfrak{X} \times_{\mathfrak{S}} \mathfrak{Y}$ exists in the category of noetherian formal schemes, see [**EGA I**], 10.13.5. Locally, one has

$$\mathrm{Spf}\,R_1 \times_{\mathrm{Spf}\,R} \mathrm{Spf}\,R_2 = \mathrm{Spf}\,R_1 \hat{\otimes}_R R_2$$

completed with respect to $I_{R_1} \otimes R_2 + R_1 \otimes I_{R_2}$.

### *Formal boundaries*

Let $\bar{X} \xrightarrow{f} S$ be a proper morphism of noetherian schemes, let $Y$ be a closed subscheme of $\bar{X}$, and let $X = \bar{X} - Y$. Assume that $\bar{X}$ is of finite type over $S$. Then in general, $X$ is not proper over $S$, and in this case, by the valuative criterion of properness, there exists a discrete valuation ring $O$ with quotient field $K$ and a commutative square

$$(1)\qquad \begin{array}{ccc} \operatorname{Spec} K & \xrightarrow{f} & X \\ \downarrow & \nearrow & \downarrow \\ \operatorname{Spec} O & \longrightarrow & S \end{array}$$

such that there exists no morphism of $\operatorname{Spec} O$ to $X$ making the whole diagram commutative, c.f. [**Ha**], exerc. II, 4.11. We can assume that $O$ is complete.

On the other hand, consider the completion $\mathfrak{X} = \bar{X}_{/Y}$ of $\bar{X}$ along $Y$. There exists a unique extension of $f$ to a morphism $\operatorname{Spec} O \xrightarrow{f} \bar{X}$. Since $f$ maps the closed point to $Y$, there exists an induced commutative square over $S$

$$(2)\qquad \begin{array}{ccc} \operatorname{Spf} O & \longrightarrow & \mathfrak{X} \\ \downarrow & \nearrow & \downarrow \\ \operatorname{Spec} O & \xrightarrow{f} & \bar{X} \end{array},$$

and there exists no morphism of $\operatorname{Spec} O$ to $\mathfrak{X}$ making the whole diagram commutative. Conversely, for any complete discrete valuation ring $O$ over $S$, any diagram of type (2) induces a diagram of type (1).

Let now $\mathfrak{X} \to S$ be an arbitrary formal scheme over a scheme $S$, and let $\operatorname{Spec} O \to S$ be the spectrum of a complete discrete valuation ring over $S$. Then we define

$$\mathfrak{X}(O)^\circ := \{g : \operatorname{Spf} O \to \mathfrak{X}/S \mid \text{there exists no extension } \operatorname{Spec} O \to \mathfrak{X}\}\,.$$

Let $X \to S$ be a separated scheme of finite type over $S$, and let $\mathfrak{X}$ be a separated formal scheme of finite type over $S$. Then $\mathfrak{X}$ is called a *formal boundary* of $X$, if for any $\operatorname{Spec} O$ over $S$, there exists a bijection, functorial in the category of complete discrete valuation rings $O$ over $S$, between $\mathfrak{X}(O)^\circ$ and the set of morphisms $\operatorname{Spec} K \to X$ over $S$ which can not be extended to a morphism $\operatorname{Spec} O \to X$.

Thus in the situation $X \xrightarrow{u} \bar{X} \xleftarrow{v} \mathfrak{X} = \bar{X}_{/Y}$ as above, $\mathfrak{X}$ is a formal boundary of $X$. Moreover, we can define a ringed space

$$< \mathfrak{X} > := < \bar{X} > := (\bar{X}, u_*\mathcal{O}_X \otimes_{\mathcal{O}_{\bar{X}}} v_*\mathcal{O}_{\mathfrak{X}})\,.$$

There is a canonical morphism of ringed spaces $< \mathfrak{X} > \longrightarrow (\bar{X}, v_*\mathcal{O}_{\mathfrak{X}})$.

LEMMA 1.3. *Let $\mathfrak{X}$ be a noetherian non discrete formal scheme. There exists a complete discrete valuation ring $O$, such that $\mathfrak{X}(O)^\circ \neq \emptyset$.*

PROOF. We may assume that $\mathfrak{X} = \operatorname{Spf}(R, I)$ is affine, where $I$ is not nilpotent. Then $IR_{\mathfrak{m}}$ is not nilpotent for some prime ideal $\mathfrak{m}$. In particular, $\dim R_{\mathfrak{m}} \geq 1$. By the lemma below, there exists a prime $\mathfrak{p} \in \operatorname{Spec} R_{\mathfrak{m}}$ of codimension 1, which does not contain $I$. Then $I$ is not zero in $R_{\mathfrak{m}}/\mathfrak{p}$. Let $O$ be the completion of the normalization of $R_{\mathfrak{m}}/\mathfrak{p}$. By the Krull-Akizuki theorem, $O$ is a complete discrete valuation ring, see [**Bou**], VII, 2.5, and the obvious morphism $\operatorname{Spf} O \xrightarrow{f} \operatorname{Spf} R$ can not be extended to $\operatorname{Spec} O$ since the inverse image $f^{\sharp-1}(0)$ is not open in $R$. □

LEMMA 1.4. *Let $R$ be a local noetherian ring of dimension $n \geq 1$, and let $I \subseteq R$ be an ideal, which is not nilpotent. Then there exists some $\mathfrak{p} \in \operatorname{Spec} R$ of codimension 1, which does not contain $I$.*

PROOF. One reduces to the case that $R$ is a local integral domain and $I$ is a principal ideal. If there exists any prime ideal $(0) \neq \mathfrak{q}$ not containing $I$, then $R/\mathfrak{q}$ has dimension less than $n$, so we can obtain the result by induction over $n$. Assume that $\dim R \geq 2$ and $I$ is contained in all prime ideals except $(0)$. Since $I$ is principal, there exists an element $a \in \mathfrak{m}_R$ not contained in $\bigcup_{\mathfrak{q} \in \operatorname{Min} I} \mathfrak{q}$, where $\operatorname{Min} I$ is the set of minimal primes over $I$. Then any minimal prime over $(a)$ does not contain $I$. This is a contradiction. □

## 2. Good and stable reduction

¿From now on, $O$ denotes a complete discrete valuation ring over $A$, $K$ its quotient field, and $v$ the valuation of $K$. By $K^s$ we denote a separable closure of $K$. The valuation extends to $K^s$ with values in $\mathbb{Q}$. Finally, we define $|a| = q^{-v(a)}$ for $a \in K^s$. Let $\mathbf{F}$ be a Drinfeld module over $K$. We say that $\mathbf{F}$ has *good reduction*, if there exists a Drinfeld module $\mathbf{E}$ over $O$ such that $\mathbf{F} \cong K \otimes \mathbf{E}$.

A Drinfeld module $\mathbf{F}$ is said to have *coefficients in* $O$, if $f_a \in O\{\tau_q\}$ for all $a \in A$. It is said to have *stable reduction*, if it is isomorphic to a Drinfeld module with coefficients in $O$ whose reduction mod $\mathfrak{m}_O$ is a Drinfeld module over $O/\mathfrak{m}_O$ (of possibly smaller rank than $\mathbf{F}$)

REMARKS 2.1. (1) It follows from 1, 4.6. that a Drinfeld module $\mathbf{F}$ has coefficients in $O$ and its reduction is a Drinfeld module if and only if there exists one non constant element $a \in A$ such that $f_a$ has coefficients in $O$ and its reduction has degree greater than 1. It has good reduction if moreover $\deg f_a = \deg(f_a \bmod \mathfrak{m}_O)$.

(2) Clearly, any Drinfeld module over $K$ is isomorphic to one with coefficients in $O$.

PROPOSITION 2.2. (i) *Let* $\mathbf{F}$ *be a Drinfeld module over* $K$. *There exists a finite separable extension* $K'$ *of* $K$ *such that* $\mathbf{F} \otimes K'$ *has stable reduction.*

(ii) *Let* $\varphi : \mathbf{F} \to \mathbf{G}$ *be an isogeny of Drinfeld modules over* $K$. *Then* $\mathbf{F}$ *has stable reduction if and only if* $\mathbf{G}$ *has stable reduction, and in this case, their reductions have the same rank.*

PROOF. Let $a \in A$ be not a constant, and let $f_a = \sum a_i \tau^i$, where $\tau = \tau_q$. As remarked in 2.1. (1), $\mathbf{F}$ has stable reduction if and only if there is an isomorphism $\mathbf{F} \cong \mathbf{F}'$ such that $f'_a$ is defined over $O$ and $f'^{\sharp}_a \bmod \mathfrak{m}_O$ is not linear, c.f. 1, 4.6. An isomorphism $c\tau^0$ of $\mathbb{G}_{a,K}$ provides such an $\mathbf{F}'$ if and only if

$$v(c) = \min_{i>0} \left( \frac{v(a_i)}{q^i - 1} \right) .$$

We can find such a $c$ in some extension of $K$, e.g. if we adjoin to $K$ a $(q^j - 1)$-th root of a regular parameter of $O$. Here $j$ is an index, where the minimum is taken. This proves (i).

Let $\mathbf{F}$ have stable reduction. We may assume that $\mathbf{F}$ has coefficients in $O$ and $\mathbf{F} \bmod \mathfrak{m}_O$ is a Drinfeld module. Moreover, replacing $\mathbf{G}$ by some conjugate module, we may assume that $\varphi$ is defined over $O$ and $\varphi \not\equiv 0 \bmod \mathfrak{m}_O$. Then the equation $\varphi \circ f_a = g_a \circ \varphi$ implies that $g_a$ is defined over $O$ and the degrees of the reductions of $f_a$ and $g_a$ coincide, see 1, 4.5.

Conversely, if $\mathbf{G}$ has stable reduction, then so does $\mathbf{F}$, using 2, 3.4. This completes the proof. □

Unless specified otherwise, we will always assume that a Drinfeld module with stable reduction has coefficients in $O$, and that its reduction is a Drinfeld module.

In the presence of level structures we can say more. Moreover, now we can prove the surjectivity of $M^d_I \to \operatorname{Spec} A$ asserted in 3, 4.1, (ii)). Let $\mathbf{E}$ be a standard Drinfeld module of rank $d$ over a scheme $S$. Fix a non unit $0 \neq a \in A$, and set $s = \log_q |a|_\infty$. Then we have locally

$$e_a = \sum_{i=0}^{s \cdot d} a_i \tau^i .$$

Let $N$ be a common multiple of $(q^{s\cdot i}-1,\ 1 \le i \le d)$, and set $N_i = \frac{N}{q^{s\cdot i}-1}$. Then we put

$$t_i = \frac{a_{s\cdot i}^{N_i}}{a_{s\cdot d}^{N_d}}\ ,\ 1 \le i \le d-1.$$

Note that the $t_i$'s globalize to sections in $\mathcal{O}(S)$, which only depend on the isomorphy class of $\mathbf{E}$.

In particular, for the universal Drinfeld module, let

$$j_a : M_I^d \longrightarrow \operatorname{Spec} A[T_1, \dots, T_{d-1}]$$

be the morphism defined by the $t_i$'s. Here $I$ is a non zero ideal satisfying $|V(I)| \ge 2$. For $d = 1$, $j_a$ is just the structure morphism to $\operatorname{Spec} A$.

PROPOSITION 2.3. *Let $I$ be as above, and let $(\mathbf{F}, \iota)$ be a Drinfeld module of rank $d$ and level $I$ over $K$.*

(i) *If $\mathbf{F}$ has good reduction over some algebraic extension of $K$, then it even has good reduction over $K$.*
(ii) *$\mathbf{F}$ has good reduction if and only if the corresponding morphism $\mathcal{O}(M_I^d) \to K$ maps $t_1, \dots, t_{d-1}$ into $O$.*
(iii) *The morphism $j_a$ is finite, surjective and flat.*

PROOF. Let $\mathbf{F}$ have good reduction over $K'$. Then the composite morphism $\operatorname{Spec} K' \longrightarrow \operatorname{Spec} K \longrightarrow M_I^d$ given by $\mathbf{F}$ extends to $\operatorname{Spec} O'$. Thus $\mathcal{O}(M_I^d)$ is mapped into $O' \cap K = O$. Hence $\mathbf{F}$ comes from a Drinfeld module over $O$.

If $\mathbf{F}$ has good reduction, it is isomorphic to a Drinfeld module over $O$ of level $I$. Therefore, $\mathcal{O}(M_I^d)$ is mapped into $O$. However, if it has not good reduction, it has stable reduction of rank $d_1 < d$ over an extension of $K$. By the definition of the $t_i$'s, we see that $t_{d_1}$ is not in $O$.

For affine varieties over a field, finiteness is equivalent to properness, c.f. [**Ha**], exerc. II, 4.1. and 4.6. So let us show that $j_a$ is proper. Let $O$ be any complete discrete valuation ring. Given a map $\operatorname{Spec} K \xrightarrow{\varphi} M_I^d \longrightarrow \operatorname{Spec} A[T_1, \dots, T_{d-1}]$, the composition can be extended to $\operatorname{Spec} O$ if and only if the corresponding Drinfeld module has invariants $t_i$ lying in $O$. From (ii), it has good reduction, i.e. $\varphi$ extends to $\operatorname{Spec} O$. This implies that $j_a$ is proper, c.f. [**Ha**], exerc. II, 4.11. Comparing dimensions, $j_a$ must be surjective.

It remains to check flatness. This will follow from the next lemma □

LEMMA 2.4. *Let $B$ be a noetherian ring, and let $C$ be a finite $B$-algebra. Suppose both rings are regular, i.e. all localizations are regular. Then $C$ is flat over $B$.*

PROOF. Let $\mathfrak{P} \in \operatorname{Spec} C$, and let $\mathfrak{p} = \mathfrak{P} \cap B$. Then $\hat{B}_{\mathfrak{p}} \to \hat{C}_{\mathfrak{P}}$ is a finite extension of local regular rings. Then $\hat{C}_{\mathfrak{P}}$ is free over $\hat{B}_{\mathfrak{p}}$, [**Se**], IV, prop. 22. Hence

$$\hat{C}_{\mathfrak{p}} = \prod_{\mathfrak{P} \cap B = \mathfrak{p}} \hat{C}_{\mathfrak{P}}$$

is free over $\hat{B}_{\mathfrak{p}}$, so $C_{\mathfrak{p}}$ is free over $B_{\mathfrak{p}}$ as well. This implies the lemma. □

Later we will see that all Drinfeld modules of level $I$ have stable reduction, provided $|V(I)| \geq 2$.

## 3. Lattices and Tate data

In this section we show that a Drinfeld module with stable reduction can be constructed from a Drinfeld module of smaller rank by dividing out a *lattice*. This is analogous to the construction given in 2, section 1, where we divided out a lattice in the additive group. Moreover, level structures can be included in this construction. The results of this section give a modular description of all diagrams considered in sec. 1, (1), (2) for $X = M_I^d$ and $S = \operatorname{Spec} A$. To begin with, let $\mathbf{E}$ be a Drinfeld module over $O$. Then, choosing a trivialization of its bundle we may identify $\mathbf{E}(K^s)$ with $K^s$. In particular, we are given a norm on $\mathbf{E}(K^s)$. By abuse of language, a subset $M$ of $\mathbf{E}(K^s)$ is *discrete*, if any ball in $\mathbf{E}(K^s)$ meets $M$ in at most finite many points.

DEFINITION 3.1. Let $\mathbf{E}$ be a Drinfeld module over $O$. A *parameterized lattice* of rank $d$ in $\mathbf{E}$ is an injective $A$-homomorphism $\nu : \Lambda \to \mathbf{E}(K^s)$, where $\Lambda$ is a projective $A$-module of rank $d$, whose image is discrete and invariant under the Galois group $\operatorname{Gal}(K^s/K)$. Two parameterized lattices $(\Lambda, \nu)$ and $(\Lambda', \nu')$ are *isomorphic*, if there is an isomorphism $\Lambda \xrightarrow{\Phi} \Lambda'$ satisfying $\nu = \Phi \circ \nu'$.

By a lattice we always mean a parameterized lattice. Any lattice induces a representation

$$\operatorname{Gal}(K^s/K) \to \operatorname{Aut} \Lambda \, .$$

We also introduce lattices with level structure.

DEFINITION 3.2. Let $0 \neq I \subseteq A$ be an ideal, and let $(\mathbf{E}, \iota)$ be a Drinfeld module of level $I$ over $O$. A lattice of *level* $I$ and of rank $d$ in $\mathbf{E}$ is a triple $(\Lambda,\ \nu,\ \kappa)$, where $\Lambda$ is a projective $A$-module of rank $d$, $\nu : I^{-1}\Lambda \to \mathbf{E}(K^s)$ is a lattice, such that the induced representation $\operatorname{Gal}(K^s/K) \to \operatorname{Aut}(I^{-1}\Lambda/\Lambda)$ is trivial and $\kappa$ is an isomorphism from $(I^{-1}/A)^d$ to $I^{-1}\Lambda/\Lambda$. (Here we write $I^{-1}\Lambda$ instead of $I^{-1} \otimes_A \Lambda$ for short.)

Two lattices of level $I$, $(\Lambda, \nu, \kappa)$ and $(\Lambda', \nu', \kappa')$, are *isomorphic*, if there is an isomorphism $I^{-1}\Lambda \xrightarrow{\Phi} I^{-1}\Lambda'$ of lattices satisfying $\kappa = \bar{\Phi} \circ \kappa'$ and $\nu = \Phi \circ \nu'$.

Hence any lattice of level $I$ induces a representation of the Galois group with values in the group of automorphisms of $\Lambda$, which are 1 mod$I$.

DEFINITION 3.3. Let $O$ be a complete discrete valuation ring over $A$. A *normal Tate datum* of rank $(d_1,\ d_2)$ over $O$ is a pair $(\mathbf{E},\ \Lambda)$, where $\mathbf{E}$ is a Drinfeld module of rank $d_1$ over $O$ and $\Lambda = (\Lambda,\ \nu)$ is a lattice in $\mathbf{E}$ of rank $d_2$. The sum $d = d_1 + d_2$ is called the *total rank* of $(\mathbf{E}, \Lambda, \nu)$.

Until non normal Tate data are defined at the end of this chapter, we just call them Tate data. Let $(\mathbf{E},\ \Lambda, \nu)$ and $(\mathbf{E}',\ \Lambda', \nu')$ be Tate data of the same rank. A *morphism* is a commutative square

$$\begin{array}{ccc} \Lambda & \xrightarrow{\nu} & \mathbf{E}(K^s) \\ \downarrow{\scriptstyle \Phi} & & \downarrow{\scriptstyle \varphi(K^s)} \\ \Lambda' & \xrightarrow{\nu'} & \mathbf{E}'(K^s) \end{array} \quad .$$

Note that $\Phi$ is injective, if $\varphi \neq 0$.

DEFINITION 3.4. A *Tate datum of level I* over the complete discrete valuation ring $O$ is a pair $(\mathbf{E}, \iota; \Lambda, \nu, \kappa)$ consisting of Drinfeld module of level $I$ over $O$ and a lattice of level $I$ in $\mathbf{E}$. A *morphism* is a commutative square

$$\begin{array}{ccc} I^{-1}\Lambda & \longrightarrow & \mathbf{E}(K^s) \\ \downarrow{\scriptstyle \Phi} & & \downarrow{\scriptstyle \varphi(K^s)} \\ I^{-1}\Lambda' & \longrightarrow & \mathbf{E}'(K^s) \end{array} \quad ,$$

where $\varphi$ is an isomorphism of Drinfeld modules of level $I$, and $\Phi$ is an isomorphism of lattices of level $I$.

PROPOSITION 3.5. *Let $d$ be a positive number. The category of Tate data of total rank $d$ over $O$ is equivalent to the full subcategory of all Drinfeld modules of rank $d$ over $K$ consisting of those having stable reduction.*

PROOF. Let $\mathbf{F}$ have stable reduction. Then $\mathbf{F}$ (mod $\mathfrak{m}_O^n$) is a Drinfeld module of rank $d_1 \leq d$. So there is a unique isomorphism $u_n = \tau^0 + \ldots$ to a standard Drinfeld module (1, 2.8), and the polynomials $u_n^\sharp$ converge to a power series $u \in O[[X]]$. We associate with $\mathbf{F}$ the Drinfeld module

$$e_a^\sharp = u^{-1}(f_a^\sharp(u))$$

of rank $d_1$ over $O$.

By the convergence lemma (1, 4.1), $u$ has infinite radius of convergence. Thus for any $c \in K$, $u(cX)$ is a power series in the Tate algebra $T_1 = T_{1,K}$, which has a decomposition

$$u(cX) = \omega_c \cdot h_c \ ,$$

where $\omega_c$ is a unit in $\overset{\circ}{T}_1$ and $h_c$ is a (separable) polynomial having all its roots within the ball of radius 1. Hence only finite many roots of $u$ lie in the ball $B(0, |c|)$. Since $h_1 = 1$, the roots of $u$ form a torsion free and Galois invariant submodule $\Lambda$ of $\mathbf{E}(K^s)$.

We claim that $\Lambda$ is a lattice of rank $d - d_1$. Let $a \in A$ be not a unit, whose image in $O$ does not vanish. From the commutative diagram

$$\begin{array}{ccc} \mathbf{E}(K^s) & \xrightarrow{e_a} & \mathbf{E}(K^s) \\ \downarrow u & & \downarrow u \\ \mathbf{F}(K^s) & \xrightarrow{f_a} & \mathbf{F}(K^s) \end{array}$$

we get that

$$\begin{aligned} e_a^{-1}(\Lambda)/\Lambda &= u^{-1} f_a^{-1}(0)/u^{-1}(0) \\ &\cong f_a^{-1}(0) \cong (A/a)^d \ . \end{aligned}$$

Hence from the exact sequence

$$0 \to e_a^{-1}(0) \to e_a^{-1}(\Lambda)/\Lambda \xrightarrow{a} \Lambda/a\Lambda \to 0$$

we conclude that $\Lambda/a\Lambda \cong (A/a)^{d-d_1}$.

Now we choose a ball of a suitable radius $r$ such that the map

$$B(0, r) \cap \Lambda \to \Lambda/a\Lambda$$

is surjective. Since $|a \cdot x| > |x|$ for all $0 \neq x \in \mathbf{E}(K^s)$, we see that $\Lambda$ is generated by the left side, which is a finite set. This proves the result.

Conversely, let be given a Tate datum $(\mathbf{E}, \Lambda)$ of rank $(d_1, d_2)$. Then we define a power series

$$u_\Lambda = X \cdot \prod_{\substack{\lambda \in \Lambda \\ \lambda \neq 0}} \left(1 - \frac{X}{\lambda}\right) .$$

This power series is additive, $\mathbb{F}_q$-linear, and has infinite radius of convergence, same proof as in 2, 1.2. Now we claim that by setting

$$f_a^\sharp = u_\Lambda(e_a^\sharp(u_\Lambda^{-1})) \ ,$$

we are given a Drinfeld module of rank $d$ over $K$. Let $a \neq 0$, and let the roots of $e_a^\sharp$ in $\bar{K}$ have multiplicity $q^m$. By the exact sequence described above $\Delta := e_a^{-1}(\Lambda)/\Lambda$ has cardinality $|a|_\infty^d/q^m$ and both series

$$u(e_a^\sharp) \quad \text{and} \quad \prod_{\bar{x}\in\Delta}(u - u(x))^{q^m} ,$$

where $u = u_\Lambda$, vanish precisely at the points of $e_a^{-1}(\Lambda)$, and each root has multiplicity $q^m$. For the left series this can be checked for $x = 0$, for the right series this is obvious. Hence, up to factor $c \in K^\times$, they coincide. So we have

$$f_a^\sharp = c \cdot \prod_{\bar{x}\in\Delta}(X - u(x))^{q^m} ,$$

which is a polynomial of degree $|a|_\infty^d$. The equation $\partial(f_a) = a$ follows immediately from the definition.

Next, let $\psi : \mathbf{F} \to \mathbf{F}'$ be a morphism of Drinfeld modules with stable reduction. If $\psi \neq 0$, using (1, 4.6), $\mathbf{F}$ and $\mathbf{F}'$ have Tate data $(\mathbf{E}, \Lambda)$ and$(\mathbf{E}', \Lambda')$ of the same rank, $\psi$ is defined over $O$, and $\psi \not\equiv 0 \bmod \mathfrak{m}_O$. We claim that $\varphi^\sharp := u'^{-1}(\psi^\sharp(u))$ defines an isogeny from $\mathbf{E}$ to $\mathbf{E}'$. In fact ($\varphi \bmod \mathfrak{m}_O^n$) is a homomorphism of standard Drinfeld modules, so it is standard (1, 2.7). From that the result follows. Clearly $\varphi$ induces a morphism from $\Lambda$ to $\Lambda'$.

Conversely let $0 \neq (\varphi, \Phi) : (\mathbf{E}, \Lambda) \to (\mathbf{E}', \Lambda')$ be a morphism of Tate data. We put $\psi^\sharp = u_\Lambda(\varphi^\sharp(u_{\Lambda'}'^{-1}))$ and have to show that this is a polynomial. Let the roots of $\varphi^\sharp$ have multiplicity $q^m$. We put $\Delta = \varphi^{-1}(\Lambda')/\Lambda$ and argue as above. □

This correspondence is called *Tate uniformization.* In particular, any isogeny $\mathbf{F} \xrightarrow{\psi} \mathbf{F}'$ has a *canonical factorization* given by the following factorization of its morphism of Tate data

$$(\mathbf{E};\, \Lambda,\, \nu) \overset{(\varphi,\, id)}{\longrightarrow} (\mathbf{E}';\, \Lambda,\, \varphi \circ \nu) \overset{(id,\, \Phi)}{\longrightarrow} (\mathbf{E}';\, \Lambda',\, \nu') .$$

¿From the second part of the above proof one derives the degree formula

$$\deg \psi = \deg \varphi \cdot |\Lambda'/\Phi(\Lambda)| .$$

PROPOSITION 3.6. *Let* $\mathbf{F}$ *have stable reduction with Tate datum* $(\mathbf{E};\, \Lambda,\, \nu)$*, and let* $(0) \neq I$ *be an ideal in* $A$*. Then* $\mathbf{F}/I$ *has stable reduction with Tate datum* $(\mathbf{E}/I;\, I^{-1}\Lambda,\, \nu')$*, where* $\nu'$ *is the image of* $\nu$ *under the isomorphism*

$$\mathrm{Hom}_A(\Lambda, \mathbf{E}(K^s)) \xrightarrow{\sim} \mathrm{Hom}_A(I^{-1}\Lambda, (\mathbf{E}/I)(K^s)) ,$$

*c.f. 3, 1.9. The projection* $\mathbf{F} \to \mathbf{F}/I$ *has Tate datum* $(p_I, P)$*, where* $p_I$ *is the projection, and* $P$ *is the embedding.*

PROOF. Recall that $\mathbf{F}/I$ has stable reduction (2.2). The image of $\nu$ is characterized by the property that for each $a \in I$, the diagram of decompositions of the multiplication by $a$ maps

$$\begin{array}{ccccc} \Lambda & \xrightarrow{P} & I^{-1}\Lambda & \xrightarrow{\Phi_a} & \Lambda \\ \downarrow{\scriptstyle \nu} & & \downarrow{\scriptstyle \nu'} & & \downarrow{\scriptstyle \nu} \\ \mathbf{E}(K^s) & \xrightarrow{p_I} & (\mathbf{E}/I)(K^s) & \xrightarrow{\varphi_a} & \mathbf{E}(K^s) \end{array}$$

commutes, c.f. 3, 1.10. Let $\mathbf{F}'$ be the Drinfeld module over $K$ corresponding to $(\mathbf{E}/I;\, I^{-1}\Lambda, \nu')$. Then, for each $a \in I$, there is a factorization of the multiplication by $a$ map

$$\mathbf{F} \overset{(p_I,P)}{\longrightarrow} \mathbf{F}' \to \mathbf{F}\,.$$

Hence there is an induced map $\mathbf{F}' \to \mathbf{F}/I$. Comparing degrees, we conclude that it is an isomorphism. □

Next we consider kernels in Tate uniformization. First we need some preliminaries on divisors in Tate uniformization.

Let $D = \operatorname{Div} s \subset \operatorname{Spec} K[X]$ be an effective divisor. We may assume that $s \in O[X]$ and $s \not\equiv 0 \bmod \mathfrak{m}_O O[X]$. Then, in the Tate algebra $\mathring{T}_1$ we have a unique decomposition

$$s = \omega \cdot s_0\ ,$$

where $\omega$ is a unit in $\mathring{T}_1$ and $s_0$ is a monic polynomial in $O[X]$. We call

$$D^\circ := \operatorname{Div} s_0$$

the *integral part* of $D$. A divisor satisfying $D = D^\circ$ is called *integral.* Clearly, a subdivisor of an integral divisor is integral. Let $u \in \mathring{T}_1$ with $\deg(u \bmod \mathfrak{m}_O \mathring{T}_1) > 0$. It defines a finite endomorphism of $\mathring{T}_1$. Let $D = \operatorname{Div} s_0$ be an integral divisor. Let $s_0(u) = \eta \cdot t_0$ be the Weierstrass decomposition as above. Then we set

$$u^*(D) := \operatorname{Div} t_0 \subset \operatorname{Spec} O[X]\ .$$

If $u$ defines an isomorphism of $\mathring{T}_1$, we put $u_*(D) := (u^{-1})^*(D)$.

PROPOSITION 3.7. *Let $\mathbf{F} \overset{\psi}{\to} \mathbf{F}'$ be an isogeny of rank $d$ Drinfeld modules over $K$ with stable reduction and let $(\mathbf{E};\, \Lambda) \overset{(\varphi,\Phi)}{\longrightarrow} (\mathbf{E}';\, \Lambda')$ be its morphism of Tate data. Then the kernels are related by the formula*

$$\ker \varphi = u_\Lambda^*(\ker(\psi)^\circ)\,.$$

PROOF. We have $\varphi^\sharp = u_{\Lambda'}^{-1}(\psi^\sharp(u_\Lambda))$. We write

$$u_{\Lambda'}^{-1} = \omega \cdot X ;$$

then $\omega$ is a unit in $\overset{\circ}{T}_1$, thus we get

$$\varphi^\sharp = \psi^\sharp(u_\Lambda) \cdot \omega(\psi^\sharp(u_\Lambda)) .$$

This shows that $u_\Lambda^*(\ker(\psi)^\circ) = \ker\varphi$. □

COROLLARY 3.8. *Let* **F** *be a Drinfeld module having stable reduction, and let* $(\mathbf{E}, \Lambda)$ *be a Tate datum for* **F**. *Then we have*

$$u_\Lambda^*(\mathbf{F}[I]^\circ) = \mathbf{E}[I] .$$

PROOF. Apply the proposition to $\mathbf{F} \to \mathbf{F}/I$ and use 3.6 □

REMARK. Now we turn to level structures. Let $\tilde{\iota}$ be a level $I$ structure on a Drinfeld module **F** with stable reduction and Tate datum $(\mathbf{E}, \Lambda)$. By 3, 3.3 we have

$$\mathbf{F}[I] = \sum_{x \in (I^{-1}/A)^d} \tilde{\iota}(x) .$$

Let $H := \{x \mid \tilde{\iota}(x) \in O\}$. Then $H$ is a submodule of $(I^{-1}/A)^d$, and we have

$$\mathbf{F}[I]^\circ = \sum_{x \in H} \tilde{\iota}(x) .$$

By 3.8, $H$ has cardinality $|(I^{-1}/A)|^{d_1}$. Moreover, by the same argument, $H \cap (\mathfrak{p}^{-1}/A)^d$ has cardinality $|(\mathfrak{p}^{-1}/A)|^{d_1}$ for each prime ideal containing $I$. This implies that $H \cong (I^{-1}/A)^{d_1}$. Let $(I^{-1}/A)^{d_1} \xrightarrow{h} H$ be an isomorphism. Then it is clear that $x \mapsto u_\Lambda^{-1}(\tilde{\iota} \circ h(x))$ is a level $I$ structure on **E**.

DEFINITION 3.9. A level $I$ structure on **F** is called *normal* if $H$ is the submodule generated by the first $d_1$ coordinates.

In particular, for a normal level structure, **E** has a canonical level $I$ structure. For any level structure $\tilde{\iota}$, there exists an automorphism $\alpha$ of $(I^{-1}/A)^d$ such that $\tilde{\iota} \circ \alpha$ is normal.

PROPOSITION 3.10. *Let* $(0) \neq I$ *be an ideal in* $A$ *that is contained in at least two prime ideals, and let* **F** *be a Drinfeld module over* $K$. *If* **F** *admits a level* $I$ *structure, then it has stable reduction over* $K$.

PROOF. Let **F** have stable reduction over $K' \supseteq K$. We may assume $K'/K$ is Galois. If $\mathbf{F} \otimes K'$ has even good reduction, **F** has good reduction (2.3, (i)).

In the general case, let $\tilde{\iota}$ be a level $I$ structure on $\mathbf{F}$, and let $\mathbf{F}\otimes K' \xrightarrow{c} \mathbf{F}'$, $c \in K'$, be an isomorphism to a Drinfeld module with coefficients in $O'$ and with reduction of rank $d_1$. Then $\mathbf{F}'$ is of level $I$ by means of $c \cdot \tilde{\iota}$. We may assume that $c \cdot \tilde{\iota}$ is normal. Let $(\mathbf{E}', \Lambda')$ be the Tate datum attached to $\mathbf{F}'$ by 3.5, so we have $f_a'^{\sharp} = u_{\Lambda'}(e_a'^{\sharp}(u_{\Lambda'}^{-1}))$. For any $\sigma \in \operatorname{Gal}(K'/K)$, let $\varepsilon_\sigma = {}^\sigma c \cdot c^{-1}$. Then we have

$$^\sigma f_a'^{\sharp}(X) = \varepsilon_\sigma f_a'^{\sharp}(\varepsilon_\sigma^{-1} X) .$$

The uniqueness of Tate uniformization implies that

$$^\sigma u_{\Lambda'}(X) = \varepsilon_\sigma u_{\Lambda'}(\varepsilon_\sigma^{-1} X) .$$

It follows that $u(X) := c^{-1} u_{\Lambda'}(cX)$ has coefficients in $K$. Hence $e_a^{\sharp} := u(f_a^{\sharp}(u^{-1}))$ defines a Drinfeld module $\mathbf{E}$ over $K$ of rank $d_1$, and $\mathbf{E} \otimes K'$ is isomorphic to $\mathbf{E}'$ by means of $c$. Moreover, suppressing restriction to $(I^{-1}/A)^{d_1}$, $\mathbf{E}'$ has level $I$ structure $\iota' = u_{\Lambda'}^{-1}(c \cdot \tilde{\iota})$. It follows from the above formulas that $\iota = c^{-1} \cdot \iota'$ is a level $I$ structure on $\mathbf{E} \otimes K'$, which is defined over $K$. By 2.3, $\mathbf{E}$ has good reduction, so there exists an element $c_0 \in K$ with $|c_0| = |c|$. Hence $\mathbf{F}$ has stable reduction. $\square$

PROPOSITION 3.11. *Let $d_1$, $d_2$ be positive numbers. The category of Tate data of rank $(d_1, d_2)$ over $O$ of level $I$ is equivalent to the full subcategory of all level $I$ Drinfeld modules over $K$ of rank $d_1 + d_2$ consisting of those having stable reduction of rank $d_1$ and normal level structure.*

PROOF. Let $(\mathbf{F}, \tilde{\iota})$ be the Drinfeld module of rank $d_1 + d_2$ as specified in the proposition, and let $(\mathbf{E}, \Lambda)$ be the corresponding Tate datum of rank $(d_1, d_2)$ of $\mathbf{F}$. We show that we can canonically provide this Tate datum with a level $I$ structure. As explained in the remark above, $\mathbf{E}$ has a canonical level $I$ structure, say $\iota$. Let

$$N := \sum_x \tilde{\iota}(x), \ x \in (0) \oplus (I^{-1}/A)^{d_2} .$$

Since $\ker \tilde{\iota} \subset H = \{x \mid \tilde{\iota}(x) \in O\}$ and $\mathbf{F}[I]^\circ = \sum_{x \in H} \tilde{\iota}(x)$ as explained above, $N$ is an $A$-submodule of $\mathbf{F}$ which is finite and étale over $\operatorname{Spec} K$. By 2, 3.2, $F/N$ is a Drinfeld module, and it has stable reduction by 2.2. Since $N^\circ = 0$, and taking into account degrees, $\mathbf{F}/N$ has Tate datum $(\mathbf{E}; I^{-1}\Lambda, \nu')$, such that the projection has Tate datum $(id, P)$, where $P$ is the embedding. Hence $\nu'$ is an extension of $\nu$. Identifying $I^{-1}\Lambda/\Lambda$ with its image in $\mathbf{F}(K^s)$, we see that $\tilde{\iota}_{|(I^{-1}/A)^{d_2}}$ defines a level $I$ structure $\kappa$ on $\Lambda$. So $(\mathbf{E}, \iota; \Lambda, \nu', \kappa)$ is a Tate datum of level $I$.

Conversely, given a Tate datum $(\mathbf{E}, \iota; \Lambda, \nu, \kappa)$, we define

$$\tilde{\iota} : (I^{-1}/A)^{d_1} \oplus (I^{-1}/A)^{d_2} \to \mathbf{F}(K^s)$$

by $\tilde{\iota}(x,y) = u_\Lambda(\iota(x)) + \overline{\nu} \circ \kappa(y)$, where $\mathbf{F}$ is the corresponding Drinfeld module and $\overline{\nu}$ is induced by $u_\Lambda \circ \nu$. Since $\mathrm{Gal}\,(K^s/K)$ acts trivially on $I^{-1}\Lambda/\Lambda$, $\tilde{\iota}$ takes image in $\mathbf{F}(K)$. Let $\mathfrak{p}$ be a prime ideal containing $I$, and let $\mathbf{F}[\mathfrak{p}] = \mathrm{Div}\,\varphi^\sharp$. Since $f_a^\sharp(\tilde{\iota}(y)) = 0$ for all $a \in \mathfrak{p}$ and $y \in (\mathfrak{p}^{-1}/A)^{d_2}$, it follows that $\varphi^\sharp(\tilde{\iota}(y)) = 0$. Hence we have

$$\mathbf{F}[\mathfrak{p}]^\circ + \tilde{\iota}(y) \subseteq \mathbf{F}[\mathfrak{p}]\,.$$

But these subdivisors are pairwise disjoint, so we have also

$$\sum_{y \in (\mathfrak{p}^{-1}/A)^{d_2}} (\mathbf{F}[\mathfrak{p}]^\circ + \tilde{\iota}(y)) = \sum_{(x,y) \in (\mathfrak{p}^{-1}/A)^d} \tilde{\iota}(x,y) \subseteq \mathbf{F}[\mathfrak{p}]\,.$$

Comparing degrees, we see that $\tilde{\iota}$ is a level $I$ structure on $\mathbf{F}$, which is normal by construction.

Morphisms are treated along the lines of (3.5). □

REMARK 3.12. Let $\tilde{\iota}_1$, $\tilde{\iota}_2$ be the restrictions of $\tilde{\iota}$ to $(I^{-1}/A)^{d_1}$ and $(I^{-1}/A)^{d_2}$, $\tilde{\iota}$ being normal. Then $(\mathbf{E}, \iota; \Lambda, \nu, \kappa)$ is a Tate datum for $(\mathbf{F}, \tilde{\iota})$ if and only if $(\mathbf{E}, \Lambda)$ is a Tate datum for $\mathbf{F}$ and one has the equations

$$\tilde{\iota}_1 = u_\Lambda \circ \iota \text{ and } \tilde{\iota}_2 = \bar{\nu} \circ \kappa\,.$$

Here we write $u_\Lambda$ for the map $\mathbf{E}(K^s) \to \mathbf{F}(K^s)$ defined by $u_\Lambda$, and $\bar{\nu}$ is the map induced by $u_\Lambda \circ \nu$.

COROLLARY 3.13. *If $V(I)$ contains at least two points, the category of Drinfeld modules of rank $d$ with normal level $I$ structure is equivalent to the category of Tate data of total rank $d$ and level $I$.*

This follows from the proposition and (3.10).

## 4. Group action

As before, $I$ is an ideal which is contained at least in two prime ideals. Recall that we had defined $\mathrm{GL}_d^0 = \mathrm{GL}_d(\hat{A}) \cdot \mathbb{A}_\mathrm{f}^\times$. We have seen in chap. 3, sec. 5 that the group $\overline{\mathrm{GL}}_d^0 = (\mathrm{GL}_d(\hat{A}) \cdot \mathbb{A}_\mathrm{f}^\times)/(\Gamma(I) \cdot k^\times)$ is acting on $M_I^d$. So it acts on $M_I^d(K)$. Thereby it stabilizes the set of Drinfeld modules with reduction of given rank $(d_1, d_2)$, see 2.2. Thus the subgroup preserving normal level structures acts on the set of isomorphism classes of normal Tate data over $O$, such that the canonical map to $M_I^d(K)$ becomes equivariant. We will describe this action explicitly. Let $G_{d_1}$ be the subgroup of $\mathrm{GL}_d^0$ of all matrices of the form

$$\alpha = \begin{pmatrix} \alpha_1 & \alpha' \\ 0 & \alpha_2 \end{pmatrix},$$

where $\alpha_1$ is a $d_1 \times d_1$- matrix. Let $(\mathbf{F}, \tilde{\iota})$ be a Drinfeld module of rank $d$, whose Tate datum has rank $(d_1, d_2)$ with normal level $I$ structure, and let $\alpha \in \mathrm{GL}_d(\hat{A})$. Then the integral part of $\tilde{\iota} \circ \alpha^{-1}$ is $\alpha((I^{-1}/A)^{d_1} \oplus (0))$, so $\alpha_*(\mathbf{F}, \tilde{\iota})$ has normal level structure if and only if $\alpha \bmod \Gamma(I)$ can be represented by a matrix in $G_{d_1}$. Since $\mathbb{A}_{\mathrm{f}}^{\times}$ stabilizes normal level structures, the image of $G_{d_1}$ in $\overline{\mathrm{GL}}_d^0$ is exactly the stabilizer of normal level structures. To get a clear description how this action for Tate data looks like, it is useful to treat three cases separately.

For $\mathbf{a} \in \hat{A} \cap \mathbb{A}_{\mathrm{f}}^{\times}$, we denote by $[\mathbf{a}]$ the ideal $\mathbf{a} \cdot \hat{A} \cap A$.

PROPOSITION 4.1. *Let $(\mathbf{E}, \iota; \Lambda, \nu, \kappa)$ be a Tate datum for $(\mathbf{F}, \tilde{\iota})$, and let $\alpha \in G_{d_1}$.*

(i) *If $\alpha$ is a matrix in $\mathrm{GL}_d(\hat{A})$ of the form $\begin{pmatrix} \alpha_1 & 0 \\ 0 & \alpha_2 \end{pmatrix}$, then $\alpha_*(\mathbf{F}, \tilde{\iota})$ has Tate datum $(\mathbf{E}, \iota \circ \alpha_1^{-1};\ \Lambda, \nu, \kappa \circ \alpha_2^{-1})$;*

(ii) *if $\alpha \in \mathrm{GL}_d(\hat{A})$ has the form $\begin{pmatrix} id & \alpha' \\ 0 & id \end{pmatrix}$, then $\alpha_*(\mathbf{F}, \tilde{\iota})$ has Tate datum $(\mathbf{E}, \iota; \Lambda, \nu', \kappa)$, where $\nu'(\lambda) = \nu(\lambda) - \iota \circ \alpha' \circ \kappa^{-1}(\bar{\lambda})$ for $\lambda \in I^{-1}\Lambda$;*

(iii) *if $\alpha = \mathbf{a} \cdot id$, where $\mathbf{a} \in \mathbb{A}_{\mathrm{f}}^{\times} \cap \hat{A}$ and $[\mathbf{a}] + I = (1)$, then $\alpha_*(\mathbf{F}, \tilde{\iota})$ has Tate datum $(\mathbf{a}_*(\mathbf{E}, \iota), [\mathbf{a}]^{-1};\ \Lambda, \nu', \kappa')$, where $\nu'$ is the image of $\nu$ under the canonical isomorphism*

$$\mathrm{Hom}_A(I^{-1}\Lambda, \mathbf{E}(K^s)) \xrightarrow{\sim} \mathrm{Hom}_A([\mathbf{a}]^{-1}I^{-1}\Lambda, (\mathbf{E}/[\mathbf{a}])(K^s)),$$

*c.f. 3, 1.9, and $\kappa' = \bar{P} \circ \kappa \circ \mathbf{a}^{-1}$, where $\bar{P}$ is the isomorphism $I^{-1}\Lambda/\Lambda \xrightarrow{\sim} I^{-1}[\mathbf{a}]^{-1}\Lambda/[\mathbf{a}]^{-1}\Lambda$ induced by the embedding.*

PROOF. Let $\tilde{\iota} = (\tilde{\iota}_1, \tilde{\iota}_2)$ as in 3.12. Then (i) follows from $\alpha_*(\mathbf{F}, \tilde{\iota}) = (\mathbf{F}, (\tilde{\iota}_1 \circ \alpha_1^{-1}, \tilde{\iota}_2 \circ \alpha_2^{-1}))$ and the equations

$$\tilde{\iota}_1 \circ \alpha_1^{-1} = u_\Lambda(\iota \circ \alpha_1^{-1})$$

and

$$\tilde{\iota}_2 \circ \alpha_2^{-1} = \bar{\nu} \circ \kappa \circ \alpha_2^{-1}.$$

In (ii) we have $\alpha_*(\mathbf{F}, \tilde{\iota}) = (\mathbf{F}, (\tilde{\iota}_1, \tilde{\iota}_2 - \tilde{\iota}_1 \circ \alpha'))$. So this case follows from the equations

$$\tilde{\iota}_1 \circ \alpha_1^{-1} = u_\Lambda(\iota \circ \alpha_1^{-1})$$

and

$$\begin{aligned} \tilde{\iota}_2 - \tilde{\iota}_1 \circ \alpha' &= \bar{\nu} \circ \kappa - u_\Lambda(\tilde{\iota} \circ \alpha') \\ &= (\bar{\nu} - u_\Lambda(\tilde{\iota} \circ \alpha' \circ \kappa^{-1}) \circ \kappa \\ &= (\overline{\nu - \iota \circ \alpha' \circ \kappa^{-1} \circ pr}) \circ \kappa. \end{aligned}$$

Let $\mathbf{F} \xrightarrow{q_{\mathbf{a}}} \mathbf{F}/[\mathbf{a}]$ and $\mathbf{E} \xrightarrow{p_{\mathbf{a}}} \mathbf{E}/[\mathbf{a}]$ be the projections, and let $\Lambda' = [\mathbf{a}]^{-1}\Lambda$. Then we have in (iii) $\alpha_*(\mathbf{F}, \tilde{\iota}) = (\mathbf{F}/[\mathbf{a}], q_{\mathbf{a}} \circ \tilde{\iota} \circ \mathbf{a}^{-1})$, c.f. 3, 5.3. By 3.6, $(\mathbf{E}/[\mathbf{a}], \Lambda', \nu'_{|\Lambda'})$ is a Tate datum for $\mathbf{E}/[\mathbf{a}]$. So the assertion follows from the equations

$$q_{\mathbf{a}} \circ \tilde{\iota}_1 \circ \mathbf{a}^{-1} = q_{\mathbf{a}} \circ u_\Lambda \circ \iota \circ \mathbf{a}^{-1} = u_{\Lambda'} \circ p_{\mathbf{a}} \circ \iota \circ \mathbf{a}^{-1}$$

and

$$q_{\mathbf{a}} \circ \tilde{\iota}_2 \circ \mathbf{a}^{-1} = q_{\mathbf{a}} \circ \bar{\nu} \circ \kappa \circ \mathbf{a}^{-1} = \bar{\nu}' \circ \bar{P} \circ \kappa \circ \mathbf{a}^{-1}.$$

This completes the proof. □

4.2. Let $\bar{G}_{d_1}$ be the image of $G_{d_1}$ in $\overline{\mathrm{GL}}_d^0$. Since the matrices considered in the proposition generate $\bar{G}_{d_1}$, we have given a complete description of the action. Let $\mathrm{NT}_I^{d_1,d_2}(O)$ be the set of isomorphism classes of normal Tate data of rank $(d_1, d_2)$ and level $I$ over $O$. Then we define the set of *Tate data* of rank $(d_1, d_2)$ and level $I$ over $O$ as the set

$$\mathrm{Tate}_I^{d_1,d_2}(O) := \overline{\mathrm{GL}}_d^0 \times^{\bar{G}_{d_1}} \mathrm{NT}_I^{d_1,d_2}(O)\,.$$

Thus, there is an equivariant bijection from this set to the set of Drinfeld modules of rank $d$ and level $I$ over $K$ with reduction of rank $d_1$.

CHAPTER 5

# Compactification of the Modular Surfaces

In this chapter we present the main result of this work. We give a detailed exposition of Drinfeld's compactification of the moduli scheme $M_I^2$. We show that this compactification is smooth, and the group action on the moduli spaces extends to their compactification.

## 1. Formal representability of Tate data

Using the results of the previous chapter, we will now construct a formal boundary of $M_I^2$. More generally, we show that Tate data of rank $(d-1,1)$ can be "represented" by a formal scheme. To begin with, we show that lattices of rank 1 with level structure are rational. From now, $O$ denotes again a complete discrete valuation ring over $A$, and $K$ its quotient field.

LEMMA 1.1. *Let $(0) \neq I$ be an ideal in $A$, let $\mathbf{E}$ be a Drinfeld module of level $I$ over $O$, and let $I^{-1}\Lambda \xrightarrow{\nu} \mathbf{E}(K^s)$ be a lattice of rank 1 and level $I$ in $\mathbf{E}$. Then $\Lambda$ is rational, i.e. $\nu(I^{-1}\Lambda) \subseteq \mathbf{E}(K)$.*

PROOF. The image of the representation

$$\mathrm{Gal}(K^s/K) \longrightarrow \mathrm{Aut}\,\Lambda \cong A^\times = \mathbb{F}_q^\times$$

lies in $A^\times \cap (1+I) = \{1\}$. □

1.2. Let now $|V(I)| \geq 2$, let $\mathfrak{a} \subseteq A$ be an arbitrary ideal, let $d \geq 2$, and let $\mathbf{E}_{\mathfrak{a}}^{\mathrm{univ}} := \mathbf{E}^{\mathrm{univ}}/\mathfrak{a}$ be the quotient by $\mathbf{E}^{\mathrm{univ}}[\mathfrak{a}]$ of the universal Drinfeld module of rank $d-1$ and level $I$. Let $\mathbf{E}_{\mathfrak{a}}^{\mathrm{univ}} \xrightarrow{\pi} M_I^{d-1}$ be the structure morphism. By 3, 1.8, for any $A$-scheme $S$, there is a canonical bijection

$$\mathrm{Mor}_A(S, \mathbf{E}_{\mathfrak{a}}^{\mathrm{univ}}) \longleftrightarrow \{(\mathbf{E}, \iota, \nu)\},$$

where $(\mathbf{E}, \iota)$ is a Drinfeld module of rank $d-1$ and level $I$ over $S$, and $\nu$ is a homomorphism from $\mathfrak{a}$ to $\mathbf{E}(S)$. Here a morphism $\sigma$ is mapped to the triple $(\mathbf{E}, \iota, \nu)$, where $(\mathbf{E}, \iota)$ corresponds to $\pi \circ \sigma$, and where $\nu$ is represented by the induced section $S \xrightarrow{\bar{\sigma}} S \times_{\pi\circ\sigma} \mathbf{E}_{\mathfrak{a}}^{\mathrm{univ}} = \mathbf{E}/\mathfrak{a}$. $\mathrm{Mor}_A(S, \mathbf{E}_{\mathfrak{a}}^{\mathrm{univ}})$ should not be confused with $\mathbf{E}_{\mathfrak{a}}^{\mathrm{univ}}(S)$!

LEMMA 1.3. *Let* $\operatorname{Spec} K \xrightarrow{\sigma} \mathbf{E}_{\mathfrak{a}}^{\mathrm{univ}}$ *be an A-morphism. Assume that* $\pi \circ \sigma$ *can be extended to a morphism* $\operatorname{Spec} O \to M_I^{d-1}$. *Then the corresponding* $(\mathbf{E}, \iota, \nu)$ *is a (normal) Tate datum if and only if* $\sigma$ *cannot be extended to a morphism* $\operatorname{Spec} O \to \mathbf{E}_{\mathfrak{a}}^{\mathrm{univ}}$.

PROOF. By assumption, $(\mathbf{E}, \iota)$ is a Drinfeld module over $O$. Assume that there is an extension $\operatorname{Spec} O \xrightarrow{\sigma} \mathbf{E}_{\mathfrak{a}}^{\mathrm{univ}}$. Then $\nu$ is a homomorphism $\mathfrak{a} \to \mathbf{E}(O)$, so it is not a lattice. Conversely, suppose that $\sigma$ is not extendable. We can assume compatible identifications $\mathbf{E}(K) \tilde{\to} K$, $\mathbf{E}(O) \tilde{\to} O$. Because $\sigma$ can not be extended, there is $a \in \mathfrak{a}$, $a \neq 0$, such that $\sigma(a) \notin O = \mathbf{E}(O)$ (upon identification). But then $|\nu(ra)| = |\varphi_r(\nu(a))| = |\nu(a)|^{\deg \varphi(r)}$ for $r \in A$. Therefore $|\nu(ra)| > 1$ for all $r \in A$, which implies that $\nu(A \cdot a)$ is a lattice in $K$. But the index $(\mathfrak{a} : A \cdot a) < \infty$, therefore $\nu(\mathfrak{a})$ is a lattice too. □

1.4. Now we notice that there is a canonical embedding of $\mathbf{E}_{\mathfrak{a}}^{\mathrm{univ}}$ into a scheme that is proper over $M_I^{d-1}$. In fact, let $\mathbf{E}$ be any Drinfeld module over an $A$-scheme $S$, and let $\mathcal{L} = \mathcal{L}_{\mathbf{E}}$ be its line bundle.The *projective bundle* to $\mathbf{E}$ is the scheme

$$\bar{\mathbf{E}} := \operatorname{Proj} \mathrm{S}_{\mathcal{O}_S}(\mathcal{L}^{-1} \oplus \mathcal{O}_S).$$

Then $\mathbf{E}$ is an open subset of $\bar{\mathbf{E}}$ and the complement is the image of the *section at infinity* $s_\infty : S \to \bar{\mathbf{E}}$.

LEMMA 1.5. *Let* $\mathbf{E}$, $\mathbf{F}$ *be Drinfeld modules over* $S$, $T$, *respectively.*

(i) *Let* $\mathbf{E} \xrightarrow{\varphi} \mathbf{F}$ *be an isogeny over* $S \xrightarrow{\alpha} T$. *Then there exists a unique extension* $\bar{\mathbf{E}} \xrightarrow{\bar{\varphi}} \bar{\mathbf{F}}$. *It is finite and flat, and* $\bar{\varphi} \circ s_{\infty,\mathbf{E}} = s_{\infty,\mathbf{F}} \circ \alpha$;

(ii) *the addition map* $\mathbf{E} \times_S \mathbf{E} \xrightarrow{add} \mathbf{E}$ *extends to a map* $\bar{\mathbf{E}} \times_S \mathbf{E} \xrightarrow{\overline{add}} \bar{\mathbf{E}}$ *satisfying* $\overline{add} \circ (s_\infty, t) = s_\infty$ *for any* $t \in \mathbf{E}(S)$.

PROOF. Omitted. □

It follows from 1.3 that the normal Tate data of rank $(d-1, 1)$ over $O$ with lattice isomorphic to $\mathfrak{a}$ correspond bijectively to the diagrams (1) in 4, sec. 1, where $X = \mathbf{E}_{\mathfrak{a}}^{\mathrm{univ}}$ and $\bar{X} = \bar{\mathbf{E}}_{\mathfrak{a}}^{\mathrm{univ}}$. Let $\mathfrak{E}_{\mathfrak{a}}^{\mathrm{univ}}$ be the completion of $\bar{\mathbf{E}}_{\mathfrak{a}}^{\mathrm{univ}}$ along the section at infinity. Then we have morphisms of formal schemes

$$\mathbf{E}_{\mathfrak{a}}^{\mathrm{univ}} \xrightarrow{u} \bar{\mathbf{E}}_{\mathfrak{a}}^{\mathrm{univ}} \xleftarrow{v} \mathfrak{E}_{\mathfrak{a}}^{\mathrm{univ}}.$$

As we explained in the previous chapter, $\mathfrak{E}_{\mathfrak{a}}^{\mathrm{univ}}$ is a formal boundary of $\mathbf{E}_{\mathfrak{a}}^{\mathrm{univ}}$. Thus lemma 1.3 can be reformulated:

PROPOSITION 1.6. *For any complete discrete valuation ring $O$, there is a bijection from $\mathfrak{E}_{\mathfrak{a}}^{\text{univ}}(O)^{\circ}$ (see 4, sec. 1 for notation) onto the set of all triples (up to isomorphy) $(\mathbf{E}, \iota, \nu)$, where $(\mathbf{E}, \iota)$ is a level $I$ Drinfeld module of rank $d-1$ over $O$ and $\mathfrak{a} \xrightarrow{\nu} \mathbf{E}(K)$ is a lattice.* □

Next, we interpret these data canonically as Tate data of level $I$. To do this, we consider the case where $\mathfrak{a}$ is isomorphic to $I^{-1}$. Let $I^{-1} \xrightarrow{\eta} \mathfrak{a}$ be a fixed isomorphism. Then we identify the data $(\mathbf{E}, \iota, \nu)$ with the Tate datum $(\mathbf{E}, \iota;\ A, \nu \circ \eta, id_{I^{-1}/A})$ of level $I$. We set $\mathfrak{M}_I^{d-1} := \mathfrak{E}_{\mathfrak{a}}^{\text{univ}}$. Now we can state:

PROPOSITION 1.7. *For any complete discrete valuation ring $O$ there is a bijection between the set $\mathfrak{M}_I^{d-1}(O)^{\circ}$ and the isomorphism classes of normal Tate data $(\mathbf{E}, \iota;\ A, \nu, \kappa)$ of level $I$, c.f. 4, 3.4, such that $\kappa$ can be lifted to an automorphism of $I^{-1}$. For $J \subseteq I$ there is a canonical restriction map $\mathfrak{M}_J^{d-1} \to \mathfrak{M}_I^{d-1}$ over $M_J^{d-1} \to M_I^{d-1}$. Moreover, there is a canonical map $\mathfrak{M}_I^{d-1}(O)^{\circ} \to M_I^d(K)$ compatible with restriction maps.*

PROOF. Let $(\mathbf{E}, \iota;\ A, \nu, \kappa)$ be given, and let $I^{-1} \xrightarrow{\tilde{\kappa}} I^{-1}$ be a lifting of $\kappa$, then $(\mathbf{E}, \iota;\ A, \nu, \kappa) \cong (\mathbf{E}, \iota;\ A, \nu \circ \tilde{\kappa}^{-1}, id)$. On the other hand, the Tate data $(\mathbf{E}, \iota;\ A, \nu, id)$ with fixed $(\mathbf{E}, \iota)$ are pairwise non isomorphic. This shows the first statement.

Let $\mathbf{F}^{\text{univ}}$ be the universal Drinfeld module over $M_J^{d-1}$. Fix an isomorphism $J^{-1} \xrightarrow{\varepsilon} \mathfrak{b}$ to some ideal $\mathfrak{b} \subseteq A$. Interpreting $\text{Mor}_A(S, \mathbf{F}_{\mathfrak{b}}^{\text{univ}})$ as the set of triples $(\mathbf{F}, \iota, \nu)$, where $\nu$ is a morphism $J^{-1} \to \mathbf{F}(S)$, restriction to $I^{-1}$ defines a map

$$Res_{I^{-1}} : \text{Mor}_A(S, \mathbf{F}_{\mathfrak{b}}^{\text{univ}}) \to \text{Mor}_A(S, \mathbf{F}_{\mathfrak{a}}^{\text{univ}}).$$

Since the Drinfeld module $(\mathbf{F}, \iota) : S \to M_J^{d-1}$ is the same, $Res$ induces a map

$$Res_{I^{-1}} : \text{Mor}_{M_J^{d-1}}(S, \mathbf{F}_{\mathfrak{b}}^{\text{univ}}) \to \text{Mor}_{M_J^{d-1}}(S, \mathbf{F}_{\mathfrak{a}}^{\text{univ}}),$$

in particular a canonical morphism $res_{I^{-1}} : \mathbf{F}_{\mathfrak{b}}^{\text{univ}} \to \mathbf{F}_{\mathfrak{a}}^{\text{univ}}$ over $M_J^{d-1}$. Since restriction of $\nu$ to $I^{-1}$ is an $A$-homomorphism, $res$ is a homomorphism of Drinfeld modules. Finally, let $0 \neq a \in A$ such that $aJ^{-1} \subseteq I^{-1}$. Taking $\varepsilon' := a^{-1}\varepsilon : aJ^{-1} \to \mathfrak{a}$, restriction of $\nu$ to $aJ^{-1}$ is represented by $\mathbf{F}_{\mathfrak{b}}^{\text{univ}} \xrightarrow{f_a} \mathbf{F}_{\mathfrak{b}}^{\text{univ}}$. Hence $res$ is an isogeny. Composing with $\mathbf{F}_{\mathfrak{a}}^{\text{univ}} \to \mathbf{E}_{\mathfrak{a}}^{\text{univ}}$ coming from restriction of level structures, we get a morphism $\mathbf{F}_{\mathfrak{b}}^{\text{univ}} \to \mathbf{E}_{\mathfrak{a}}^{\text{univ}}$ over $M_J^{d-1} \to M_I^{d-1}$, which extends to $\bar{\mathbf{F}}_{\mathfrak{b}}^{\text{univ}} \to \bar{\mathbf{E}}_{\mathfrak{a}}^{\text{univ}}$, c.f. 1.5. The required restriction map is obtained by completion along $s_\infty$. The map $\mathfrak{M}_I^{d-1}(O)^{\circ} \to M_I^d(K)$ is defined by taking the associated Drinfeld module over $K$. □

The construction of Proposition 1.7. gives us, for any $O$, an embedding

$$\mathfrak{N}_I^{d-1}(O)^\circ \to \mathrm{NT}_I^{d-1,1}(O) \subseteq \mathrm{Tate}_I^{d-1,1}(O).$$

We identify $\mathfrak{N}_I^{d-1}(O)$ with its image. To cover the full set of Tate data, we use the $\overline{\mathrm{GL}}_d^0$-action on them established in chapter 4, sec. 4. To this end, we first determine the subgroup which stabilizes the subset $\mathfrak{N}_I^{d-1}(O)^\circ$. Moreover, we show that this subgroup actually acts on $\mathfrak{N}_I^{d-1}$.

LEMMA 1.8. *Let* $\mathbf{E} = (\mathbf{E}, \iota;\, A, \nu, id)$ *be in* $\mathfrak{N}_I^{d-1}(O)^\circ$, *and let* $\alpha \in \overline{\mathrm{GL}}_d^0$. *Then* $\alpha_*\mathbf{E}$ *is in* $\mathfrak{N}_I^{d-1}(O)^\circ$ *if and only if* $\alpha$ *can be represented by a matrix of the form* $\begin{pmatrix}\alpha_1 & \alpha' \\ 0 & 1\end{pmatrix}$, *where* $\alpha_1 \in \mathrm{GL}_{d-1}(\hat{A})$ *and* $\alpha' \in \hat{A}^{d-1}$. *Moreover, let* $\bar{G}$ *be the image of these matrices in* $\overline{\mathrm{GL}}_d^0$. *There is a* $\bar{G}$*-action on* $\mathfrak{N}_I^{d-1}$, *which induces the* $\bar{G}$*-action on* $\mathfrak{N}_I^{d-1}(O)^\circ$ *for any* $O$.

PROOF. Let $\alpha_*\mathbf{E} \in \mathfrak{N}_I^{d-1}(O)^\circ$. By 4, sec. 4, $\alpha$ can be represented by a matrix in $G_{d_1}$. If $\alpha$ has the form $\begin{pmatrix}\alpha_1 & \alpha' \\ 0 & 1\end{pmatrix}$, then $\alpha_*\mathbf{E} \in \mathfrak{N}_I^{d-1}(O)^\circ$ by 4, 4.1, (i), (ii). On the other hand, let $\alpha = \mathbf{a}\cdot id$ with $\mathbf{a} \in \mathbb{A}_f^\times \cap \hat{A}$ and $[\mathbf{a}] + I = (1)$. Then multiplication by $\mathbf{a}$ on $I^{-1}/A$ can be lifted to an automorphism of $I^{-1}$ if and only if $\mathbf{a} \in (1 + I\hat{A})\cdot \mathbb{F}_q^\times$. This proves the first assertion.

Let $\alpha \in \bar{G}$ be represented by $\begin{pmatrix}\alpha_1 & \alpha' \\ 0 & 1\end{pmatrix}$. Then $\alpha$ determines an automorphism

$$(I^{-1}/A)^{d-1} \oplus I^{-1} \xrightarrow{\begin{pmatrix}\alpha_1 & \alpha' \\ 0 & 1\end{pmatrix}} (I^{-1}/A)^{d-1} \oplus I^{-1}$$

which is independent from the representative. Hence for any $A$-scheme $S$, $\bar{G}$ acts on triples $(\mathbf{F},\ \iota\ , \nu)$ over $S$ by

$$\alpha_*(\mathbf{F},\ \iota\ , \nu) = (\mathbf{F},\ \iota_\alpha\ , \nu_\alpha)\,,$$

where $\iota_\alpha + \nu_\alpha = (\iota + \nu)\circ \alpha^{-1}$. Here $\iota + \nu$ is the homomorphism from $(I^{-1}/A)^{d-1} \oplus I^{-1}$ to $\mathbf{F}(S)$. This defines an action of $\bar{G}$ on $\mathbf{E}_{\mathfrak{a}}^{\mathrm{univ}}$. Since $\iota_\alpha = \iota\circ\alpha_1^{-1}$, it is an action over the action of $\mathrm{GL}_{d-1}(\hat{A})$ on $M_I^{d-1}$. If $\alpha = \begin{pmatrix}\alpha_1 & 0 \\ 0 & 1\end{pmatrix}$, this is just the action of $\alpha_1$ on $\mathbf{E}^{\mathrm{univ}}$, see 3, 5.4. On the other hand, if $\alpha = \begin{pmatrix}1 & \alpha' \\ 0 & 1\end{pmatrix}$, $\alpha'$ defines an $A$-homomorphism

$I^{-1} \xrightarrow{\alpha'} I^{-1}/A$. Hence $\iota \circ \alpha'$ is represented by a section $M_I^{d-1} \xrightarrow{\sigma} \mathbf{E}_{\mathfrak{a}}^{\mathrm{univ}}[I]$. Then $\alpha$ acts by $id_{\mathbf{E}_{\mathfrak{a}}^{\mathrm{univ}}} - \sigma \circ \pi$. This describes our action explictly. We see that $\bar{G}$ is acting on $\mathbf{E}_{\mathfrak{a}}^{\mathrm{univ}}$ by means of affine morphisms.

By 1.5, the action of $\bar{G}$ extends to an action on $\bar{\mathbf{E}}_{\mathfrak{a}}^{\mathrm{univ}}$ leaving invariant the section at infinity. Hence it induces an action on $\mathfrak{E}_{\mathfrak{a}}^{\mathrm{univ}} = \mathfrak{N}_I^{d-1}$.

Now let $O$ be a complete discrete valuation ring with quotient field $K$. Comparing the action on $M_I^d$, see 5.3, one verifies that the embedding of $\mathfrak{N}_I^{d-1}(O)^\circ$ into $M_I^d(K)$, c.f. 1.7, is $\bar{G}$-equivariant. This proves the last statement, c.f. 4, sec. 4. □

LEMMA 1.9. *The group $\overline{\mathrm{GL}_d^0}$ is finite.*

PROOF. Consider the exact sequence

$$1 \to \mathrm{GL}_d(\hat{A})/(\Gamma(I) \cdot (k^\times \cap \mathrm{GL}_d(\hat{A}))) \to \overline{\mathrm{GL}_d^0} \to \mathrm{GL}_d^0/k^\times \cdot \mathrm{GL}_d(\hat{A}) \to 1\,.$$

Here the left hand group is isomorphic to $\mathrm{GL}_d(A/I)/\mathbb{F}_q^\times$, so it is finite.

On the other hand, the right hand group is isomorphic to $\mathbb{A}_{\mathrm{f}}^\times/k^\times \cdot \hat{A}^\times$. But $\mathbb{A}_{\mathrm{f}}^\times/\hat{A}^\times$ is the set of divisors $D$ on $\mathcal{C} - \{\infty\} = \operatorname{Spec} A$. For each such divisor there is a unique integer $n$, such that the divisor $D' = D + n \cdot \infty$ on $\mathcal{C}$ of degree lying in $\{0, \ldots r-1\}$, where $r$ is the degree of the residue field $\kappa(\infty)$ over $k_0$. Therefore, as a set, the left hand group can be identified with the classes of such divisors, thus it has cardinality $r \cdot \mathrm{Cl}(k)$ (in fact, there are divisors of any degree, c.f. [**Weil**], VII.5, Cor. 5). It is well known that the class number $\mathrm{Cl}(k)$ is finite, c.f. [**Hasse**], p. 570f. □

Now we can define a formal scheme that will cover all Tate data. Let $\bar{G}$ be as in (1.8). Then we define

$$\mathfrak{M}_I^{d-1,1} := \overline{\mathrm{GL}_d^0} \times^{\bar{G}} \mathfrak{N}_I^{d-1}\,.$$

It follows from (1.9) that $\mathfrak{M}_I^{d-1,1}$ is a finite disjoint union of copies of $\mathfrak{N}_I^{d-1}$, see appendix A. Similarly, we define schemes

$$M_I^{d-1,1} := \overline{\mathrm{GL}_d^0} \times^{\bar{G}} \mathbf{E}_{\mathfrak{a}}^{\mathrm{univ}}$$

and

$$\bar{M}_I^{d-1,1} := \overline{\mathrm{GL}_d^0} \times^{\bar{G}} \bar{\mathbf{E}}_{\mathfrak{a}}^{\mathrm{univ}}\,.$$

Then we have canonical morphisms

$$M_I^{d-1,1} \xrightarrow{\rho} \bar{M}_I^{d-1,1} \xleftarrow{\hat{\rho}} \mathfrak{M}_I^{d-1,1}\,,$$

where the right hand side is the completion along the complement of the left hand side.

THEOREM 1.10. *For any complete discrete valuation ring $O$, there is a canonical bijection*

$$\varphi_I : \mathfrak{M}_I^{d-1,1}(O)^\circ \to \mathrm{Tate}_I^{d-1,1}(O)\,.$$

*It induces a $\overline{\mathrm{GL}}_d^0$-equivariant map $\Phi_I : \mathfrak{M}_I^{d-1,1}(O)^\circ \to M_I^d(K)$, which is a bijection onto the set of Drinfeld modules with bad reduction of rank $d-1$. In particular, for $d = 2$, $\mathfrak{M}_I^{1,1}$ is a formal boundary of $M_I^2$.*

PROOF. $\varphi_I$ is injective by (1.8). To prove surjectivity, it is enough to show that $\mathrm{NT}_I^{d-1,1}(O)$ is contained in the image. So let $(\mathbf{E},\, \iota;\, \Gamma,\, \nu,\, \kappa)$ be a normal Tate datum of rank $(d-1, 1)$ and level $I$ over $O$. Choose $\mathbf{a} \in \hat{A}$ such that $\Gamma \cong [\mathbf{a}]$, and put $\alpha = \mathbf{a} \cdot id$. Then

$$\alpha_* \mathbf{E} = ((\alpha_1)_*(\mathbf{E},\, \iota);\, A,\, \nu',\, \kappa')\,,$$

where $\kappa' : I^{-1}/A \tilde{\to} I^{-1}/A$ is a suitable isomorphism, which can be represented by some element $\beta_2 \in \hat{A}^\times$.

Let $\beta := \begin{pmatrix} 1 & 0 \\ 0 & \beta_2 \end{pmatrix}$. Then $(\beta\alpha)_* \mathbf{E}$ has $\kappa = id_{I^{-1}/A}$, c.f. 4, 4.1. Thus it is a point in $\mathfrak{N}_I^{d-1}(O)^\circ$ by 1.7. The rest follows from the group theoretic description of Tate data in 4, 4.2. □

COROLLARY 1.11. *Let $x$ be a point in* $\operatorname{Spec} A - V(I)$*, let $\kappa(x)$ be its residue field, and let $\bar{C}$ be the smooth compactification of the smooth curve $C = M_I^2 \times_A \kappa(x)$ over* $\operatorname{Spec} \kappa(x)$*. Then the completion $\hat{C}$ of $\bar{C}$ along $\bar{C} - C$ is isomorphic to $\mathfrak{M}_I^{1,1} \times_A \kappa(x)$.*

PROOF. Both, $\hat{C}$ and $\mathfrak{M}_I^{1,1} \times \kappa(x)$ are formal boundaries of $C$ consisting of a disjoint union of spectra of complete discrete valuation rings. Therefore by the defining property of a formal boundary, $\hat{C}$ and $\mathfrak{M}_I^{1,1} \times \kappa(x)$ are isomorphic. □

Finally, let $J \subseteq I$. Then there is a canonical restriction map

$$\mathfrak{M}_J^{d-1,1} \xrightarrow{r} \mathfrak{M}_I^{d-1,1}\,.$$

We show that this is the quotient by the action of $\Gamma_{I,J} \subseteq \mathrm{GL}_d(\hat{A})/\Gamma(J)$, see 3, 5.2 for notation.

PROPOSITION 1.12. *The canonical map $\Gamma_{I,J} \backslash \mathfrak{M}_J^{d-1,1} \to \mathfrak{M}_I^{d-1,1}$ is an isomorphism.*

PROOF. We have to show that $\mathcal{O}_{\mathfrak{M}_I} = (r_* \mathcal{O}_{\mathfrak{M}_J})^{\Gamma_{I,J}}$. Since $\Gamma_{I,J}$ is a normal subgroup of $\overline{\mathrm{GL}}^0_{d,J} := (\mathrm{GL}_d(\hat{A}) \cdot \mathbb{A}_f^\times)/(\Gamma(J) \cdot k^\times)$, we have to show that $\mathfrak{N}_I^{d-1} = \Gamma_{I,J} \cap \bar{G} \backslash \mathfrak{N}_J^{d-1}$, where $\bar{G}$ is the group defined in 1.8. We will use the notation of 1.7. As in 3, 5.2 one checks that the restriction map

$$\mathbf{F}_{\mathfrak{b}}^{\mathrm{univ}} \times_{\operatorname{Spec} A} (\operatorname{Spec} A - V(J)) \to \mathbf{E}_{\mathfrak{a}}^{\mathrm{univ}} \times_{\operatorname{Spec} A} (\operatorname{Spec} A - V(J))$$

makes the left space to a principal homogeneous space with group $\Gamma_{I,J} \cap \bar{G}$. By the same normality argument we conclude that $\mathbf{E}_{\mathfrak{a}}^{\mathrm{univ}} = \Gamma_{I,J} \cap \bar{G} \backslash \mathbf{F}_{\mathfrak{b}}^{\mathrm{univ}}$. Since $\bar{\mathbf{F}}_{\mathfrak{b}}^{\mathrm{univ}} \backslash \mathbf{F}_{\mathfrak{b}}^{\mathrm{univ}}[J]$ is an affine open subset of $\bar{\mathbf{F}}_{\mathfrak{b}}^{\mathrm{univ}}$, which is $\Gamma_{I,J}$-invariant, we also get $\bar{\mathbf{E}}_{\mathfrak{a}}^{\mathrm{univ}} = \Gamma_{I,J} \cap \bar{G} \backslash \bar{\mathbf{F}}_{\mathfrak{b}}^{\mathrm{univ}}$. Let $\mathcal{I}_\infty$ be the ideal of the section at infinity in $\bar{\mathbf{E}}_{\mathfrak{a}}^{\mathrm{univ}}$. It generates an ideal of definition in $\mathcal{O}_{\mathfrak{N}_J^{d-1}}$. Since completion with respect to $\mathcal{I}_\infty$ commutes with taking invariants, we are done. □

## 2. The universal Drinfeld module with bad reduction

In this section we construct a "universal Drinfeld module" on the formal boundary $\mathfrak{M} = \mathfrak{M}_I^{d-1,1}$ . This is an $A$-module structure on an invertible $\mathcal{O}_{\mathfrak{M}}$-sheaf, that induces a pseudo-Drinfeld module on $< \mathfrak{M} >$ of rank $d$ and level $I$. Its universality amounts to the fact that over any complete discrete valuation ring, any Drinfeld module with bad reduction of rank $(d-1, 1)$ arises from it by base change.

We begin with a local construction. Let $(0) \neq I$ be an ideal in $A$ with $|V(I)| \geq 2$, and let $I^{-1} \xrightarrow{\eta} \mathfrak{a}$ be a fixed isomorphism onto an ideal of $A$. Let $R$ be an $A$-algebra, and let $(\mathbf{E}, \iota)$ be a Drinfeld module over $\operatorname{Spec} R$ of rank $d-1$ and level $I$ with trivial bundle. Hence $\mathbf{E} \cong \operatorname{Spec} R[Z]$. For $a \in I^{-1}$, let $\mathbf{E} \xrightarrow{p_a} \mathbf{E}/\mathfrak{a} \xrightarrow{\varphi_a} \mathbf{E}$ be the canonical factorization of $e_{\eta(a)}$, see 3, 1.7.

DEFINITION 2.1. Identifying $\mathbf{E}/\mathfrak{a}\,(R((1/X)))$ with $R((1/X))$, the *universal lattice* of type $I^{-1}$ is the homomorphism $I^{-1} \xrightarrow{\nu} \mathbf{E}\,(R((1/X)))$, such that the corresponding section in $(\mathbf{E}/\mathfrak{a})\,(R((1/X)))$ is $X$, see 3, 1.8.

2.2. Now we imitate the construction of a Drinfeld module of rank $d$ with bad reduction, see 4, 3.5. Setting $\Lambda := \nu(A)$, we define a power series

$$u_\Lambda := Z \cdot \prod_{\lambda \in \Lambda} \left(1 - \frac{Z}{\lambda}\right) = Z \cdot \prod_{a \in A} \left(1 - \frac{Z}{\varphi_a^\sharp(X)}\right).$$

This series has the following properties

(i) $1/\varphi_a^\sharp(X) \in R[[1/X]]$, and therefore $u_\Lambda \in R[[1/X]][[Z]]$;

(ii) $u_\Lambda$ is $\mathbb{F}_q$-linear in $Z$, and $u_\Lambda \equiv Z \bmod (1/X)$;
(iii) $u_\Lambda$ has infinite radius of convergence, i.e. for all $N \in \mathbb{N}$, writing $u_\Lambda = \sum u_n(1/X)Z^n$, the $u_n$ satisfy $u_n(1/X) \cdot X^{Nn} \in R[[1/X]]$ for almost all $n$.

For (ii) observe that for $a \neq b$ in $A$, $\varphi_a^\sharp(X) - \varphi_b^\sharp(X) = \varphi_{a-b}^\sharp(X)$ is a non zero divisor in $R[X]$. Therefore the arguments of 2, 1.2 do work, even though $R$ may have zero divisors.

Now, for $a \in A$, we set

$$t_a^\sharp := u_\Lambda(e_a^\sharp(u_\Lambda^{-1}))\,.$$

Then we have

PROPOSITION 2.3. (i) *$t_a^\sharp$ is a polynomial in $R[[1/X]][Z]$ of degree $|a|_\infty^d$;*
(ii) *after inverting $1/X$, the family $(t_a^\sharp)_{a\in A}$ defines a Drinfeld module of rank $d$ over $R((1/X))$;*
(iii) *this Drinfeld module has a canonical level $I$ structure.*

PROOF. First we prove this for restrictions to suitable open subsets of $M_I^{d-1}$ of the universal Drinfeld module $(\mathbf{E}^{\text{univ}}, \iota)$. All assertions are local in the base, so it will be sufficient to prove them for $\mathbf{E}^{\text{univ}} \otimes R$, where $R = \mathcal{O}_{M,m}$ for a closed point $m$ of $M$. Since $R$ is an integral domain, (i) can be verified over its quotient field $L$. But in this case the assertion follows from 4, 3.5, applied to $\mathbf{E}^{\text{univ}} \otimes_{\mathcal{O}_M} L[[1/X]]$ and the lattice generated by $\varphi_1^\sharp(X)$.

To prove (ii), we have to show that the leading coefficient of $t_a^\sharp$ is a unit in $R((1/X))$. Let

$$c\left(\frac{1}{X}\right) = \sum_{\nu=e}^{\infty} c_\nu \cdot \left(\frac{1}{X}\right)^\nu ,$$

$c_e \neq 0$, be this coefficient. Then we must show that $c_e \in R^\times$.

Since $R$ is a normal domain, we have $\bigcap_{\text{ht}\mathfrak{p}=1} R_\mathfrak{p} = R$. Thus it will be sufficient to prove $c_e \in \hat{R}_\mathfrak{p}^\times$ for all primes of height 1. Or, equivalentely,

$$c\left(\frac{1}{X}\right) \in \hat{R}_\mathfrak{p}\left(\left(\frac{1}{X}\right)\right)^\times .$$

By the Weierstrass preparation theorem, $\hat{R}_\mathfrak{p}[[1/X]]$ is factorial, so $c(1/X)$ can be decomposed uniqly into a product

$$c\left(\frac{1}{X}\right) = \pi^m \cdot \left(\frac{1}{X}\right)^n \cdot P_1 \dots P_r \cdot \omega\,,$$

where $\pi \cdot \hat{R}_{\mathfrak{p}} = \mathfrak{p}\hat{R}_{\mathfrak{p}}$, $\omega \in \hat{R}_{\mathfrak{p}}[[1/X]]^{\times}$, and $P_1, \ldots, P_r$ are irreducible Weierstrass polynomials not associated to $1/X$. These are prime in $\hat{R}_{\mathfrak{p}}[[1/X]]$.

If $m > 0$, the polynomial $\bar{t}_a^{\sharp} := t_a^{\sharp} \bmod \mathfrak{p}\hat{R}_{\mathfrak{p}}$ has a degree less than $|a|_{\infty}^d$. This is not possible by 4, 3.5, since $X$ generates a lattice in $(\hat{R}_{\mathfrak{p}}/\mathfrak{p}\hat{R}_{\mathfrak{p}})((1/X)) = Q(\hat{R}_{\mathfrak{p}}/\mathfrak{p}\hat{R}_{\mathfrak{p}}[[1/X]])$. Therefore we have $m = 0$.

Assuming now $r > 0$, we consider $\bar{R} := \hat{R}_{\mathfrak{p}}[[1/X]]/P_1$. This is a domain and a finite $\hat{R}_{\mathfrak{p}}$-module. The valuation on $Q(\hat{R}_{\mathfrak{p}})$ extends uniquely to a valuation on $Q(\bar{R})$. Let $O$ be the valuation ring. Since $P_1$ is Weierstrass and $\pi \in \mathfrak{m}_O$, the image of $1/X$ is not a unit in $O$. Hence $X$ generates a lattice in $Q(O) = Q(\bar{R})$. Again, $t_a^{\sharp} \bmod (P_1)$ must be a polynomial of degree $|a|_{\infty}^d$. This is a contradiction. Hence $c(1/X) = (1/X)^n \cdot \omega$ is a unit in $\hat{R}_{\mathfrak{p}}((1/X))$ as desired. Denote this Drinfeld module by $\mathbf{T}'$.

Now we define a level $I$ structure on $\mathbf{T}'$. We set

$$\tilde{\iota}(x, \bar{y}) = u_{\Lambda}\left(\frac{1}{X}, \iota(x) + \varphi_y^{\sharp}(X)\right) \in R\left(\left(\frac{1}{X}\right)\right)$$

for $x \in (I^{-1}/A)^{d-1}$ and $y \in I^{-1}$. This is well defined and additive by 2.2, (ii) and (iii). Moreover,

$$\tilde{\iota}(ax, a\bar{y}) = u_{\Lambda}\left(\frac{1}{X}, e_a^{\sharp}(\iota(x) + \varphi_y^{\sharp}(X))\right) = t_a^{\sharp}\left(u_{\Lambda}\left(\frac{1}{X}, \iota(x) + \varphi_y^{\sharp}\right)\right),$$

so $\tilde{\iota}$ is $A$-linear. To show that it is a level $I$ structure, we may pass to $L((1/X))$. But then the result follows from 4, 3.11, since $(\mathbf{E}^{\mathrm{univ}} \otimes L[[1/X]], \iota \otimes id; \nu, id)$ is a Tate datum of level $I$ inducing $\tilde{\iota}$.

Now we can pass to the general case. Again, it will be sufficient to prove the statements if $R$ is any local ring. In this case, $(\mathbf{E}, \iota)$ and the $t_a^{\sharp}$'s arise from the special case considered above by base change. In virtue of (ii), the degree of of $t_a^{\sharp}$ will not decrease by any base change. □

We put this construction in a more geometric framework. We identify $\operatorname{Spf} R[[1/X]]$ with the completion of $\overline{\mathbf{E}/\mathfrak{a}}$ along the section at infinity. We denote it by $\mathfrak{E}/\mathfrak{a}$. There are canonical mappings $\mathbf{E}/\mathfrak{a} \to \overline{\mathbf{E}/\mathfrak{a}} \leftarrow \mathfrak{E}/\mathfrak{a}$, and the $t_a^{\sharp}$ define a homomorphism

$$t_{\mathbf{E}} : A \longrightarrow \operatorname{End}^{\mathrm{bdl}}(\mathcal{O}_{\mathfrak{E}/\mathfrak{a}}[Z]),$$

c.f. 2, sec. 6. Then the second part of the proposition states that $t_{\mathbf{E}} \otimes \mathcal{O}_{<\mathfrak{E}/\mathfrak{a}>}$ is a pseudo-Drinfeld module of rank $d$ and level $I$ on $<\mathfrak{E}/\mathfrak{a}>$. In this form, the proposition can be globalized.

PROPOSITION 2.4. *Let* $(\mathbf{E}, \iota)$ *be a Drinfeld module of rank* $d-1$ *and level* $I$ *over a separated noetherian scheme* $S$. *Then the above local construction glues to a homomorphism*

$$t_{\mathbf{E}} : A \longrightarrow \mathrm{End}^{\mathrm{bdl}}(\mathrm{S}_{\mathcal{O}_{\mathfrak{E}/\mathfrak{a}}}(\mathcal{L}_{\mathbf{E}}^{-1} \otimes_{\mathcal{O}_S} \mathcal{O}_{\mathfrak{E}/\mathfrak{a}}))\,,$$

*which induces a pseudo-Drinfeld module of rank* $d$ *on* $< \mathfrak{E}/\mathfrak{a} >$. *Moreover, this pseudo-Drinfeld module has a canonical level* $I$ *structure.*

PROOF. This is an exercise in sheaf theory. Let $S = \bigcup S_\alpha$ be an open affine covering, such that $\mathcal{L}_{\mathbf{E}}$ is trivial over $S_\alpha$. Let $\mathcal{O}_{S_\alpha} \stackrel{\varphi_\alpha}{\to} (\mathcal{L}_{\mathbf{E}})_{|S_\alpha}$ and $\mathcal{O}_{S_\alpha} \stackrel{\psi_\alpha}{\to} (\mathcal{L}_{\mathbf{E}/\mathfrak{a}})_{|S_\alpha}$ be trivializations, c.f. 1, 3.2, and let $(\lambda_{\alpha\beta}) = (\varphi_\alpha^{-1}\varphi_\beta)$ and $(\mu_{\alpha\beta}) = (\psi_\alpha^{-1}\psi_\beta)$ be the corresponding cocycles in $\mathcal{O}_S^\times$. Let $R_\alpha := \mathcal{O}_S(S_\alpha)$. Then we can identify $\mathbf{E}_{|S_\alpha}$ and $\mathbf{E}/\mathfrak{a}_{|S_\alpha}$ with $\operatorname{Spec} R_\alpha[Z_\alpha]$ and $\operatorname{Spec} R_\alpha[X_\alpha]$ in such a way that we have $Z_\alpha = \lambda_{\alpha\beta}Z_\beta$ and $X_\alpha = \mu_{\alpha\beta}X_\beta$ over $S_\alpha \cap S_\beta$. Now we must verify that

$$\frac{1}{\lambda_{\alpha\beta}} t^\sharp_{\alpha,a} = t^\sharp_{\beta,a}\left(\frac{\mu_{\alpha\beta}}{X_\alpha}, \frac{Z_\alpha}{\lambda_{\alpha\beta}}\right).$$

Here the index $\alpha$ indicates the expression in the coordinates $Z_\alpha$, $X_\alpha$. We have $\varphi^\sharp_{\alpha,a}(X_\alpha) = \lambda_{\alpha\beta}\varphi^\sharp_{\beta,a}(X_\alpha/\mu_{\alpha\beta})$ and therefore

$$u_\alpha\left(\frac{1}{X_\alpha}, Z_\alpha\right) = \lambda_{\alpha\beta} \cdot u_\beta\left(\frac{\mu_{\alpha\beta}}{X_\alpha}, \frac{Z_\alpha}{\lambda_{\alpha\beta}}\right).$$

Thus we obtain

$$\begin{aligned}
\frac{1}{\lambda_{\alpha\beta}} t^\sharp_{\alpha,a} &= \frac{1}{\lambda_{\alpha\beta}} u_\alpha\left(\frac{1}{X_\alpha}, e^\sharp_{\alpha,a}(u_\alpha^{-1})\right) \\
&= u_\beta\left(\frac{\mu_{\alpha\beta}}{X_\alpha}, \frac{1}{\lambda_{\alpha\beta}} e^\sharp_{\alpha,a}\left(\lambda_{\alpha\beta}u_\beta^{-1}\left(\frac{\mu_{\alpha\beta}}{X_\alpha}, \frac{Z_\alpha}{\lambda_{\alpha\beta}}\right)\right)\right) \\
&= u_\beta\left(\frac{\mu_{\alpha\beta}}{X_\alpha}, e^\sharp_{\beta,a}\left(u_\beta^{-1}\left(\frac{\mu_{\alpha\beta}}{X_\alpha}, \frac{Z_\alpha}{\lambda_{\alpha\beta}}\right)\right)\right) \\
&= t^\sharp_{\beta,a}\left(\frac{\mu_{\alpha\beta}}{X_\alpha}, \frac{Z_\alpha}{\lambda_{\alpha\beta}}\right)
\end{aligned}$$

as desired.

For the level structure we observe that the local level structures

$$\iota_\alpha : (I^{-1}/A)^{d-1} \to R_\alpha \text{ and } \iota_\beta : (I^{-1}/A)^{d-1} \to R_\beta$$

are identified over $S_\alpha \cap S_\beta$ by $\iota_\alpha(x) = \lambda_{\alpha\beta}\iota_\beta(x)$. Using this, one verifies that also

$$\tilde{\iota}_\alpha(x) = \lambda_{\alpha\beta}\tilde{\iota}_\beta(x)\,.$$

This is what we need to define a level structure over $< \mathfrak{E}/\mathfrak{a} >$. □

We denote this pseudo-Drinfeld module by $(\mathbf{T}'_{\mathbf{E}}, \tilde{\iota})$. Let $T \xrightarrow{\xi} S$ be a morphism of $A$-schemes, and let $\mathbf{E}' := \mathbf{E} \times_S T$. The canonical map $\bar{\mathbf{E}}' \to \mathbf{E}$ induces morphisms of locally ringed spaces $\mathfrak{E}' \to \mathfrak{E}$ and $< \mathfrak{E}' > \overset{<\xi>}{\to} < \mathfrak{E} >$ over $\xi$. Then we have

$$\mathbf{T}'_{\mathbf{E}\times T} = < \xi >^* \mathbf{T}'_{\mathbf{E}} .$$

By 2, 6.4., we have a morphism

$$\mathcal{O}(M_I^d) \xrightarrow{\vartheta_{\mathbf{E}}} \mathcal{O}(< \mathfrak{E}/\mathfrak{a} >) .$$

We will apply this to the universal Drinfeld module. We have seen that the group $\bar{G} \subseteq \overline{\mathrm{GL}}_d^0$ is acting on both, $\mathcal{O}(M_I^d)$ and $\mathcal{O}(< \mathfrak{M}_I^{d-1} >) = \mathcal{O}(< \mathfrak{E}_{\mathfrak{a}}^{\mathrm{univ}} >)$, see 3, sec. 5 and 1.8. Recall that $\bar{G}$ is the image of matrices $\begin{pmatrix} \alpha_1 & \alpha' \\ 0 & 1 \end{pmatrix}$ in $\mathrm{GL}_d(\hat{A})/\Gamma(I) \cdot k^\times \subseteq \overline{\mathrm{GL}}_d^0$.

PROPOSITION 2.5. *The map* $\vartheta_{\mathbf{E}_{\mathfrak{a}}^{\mathrm{univ}}} : \mathcal{O}(M_I^d) \to \mathcal{O}(< \mathfrak{M}_I^{d-1} >)$ *is* $\bar{G}$*-equivariant.*

PROOF. Let $\alpha \in \bar{G}$. We must show that

$$\alpha^*(\mathbf{T}', \tilde{\iota}) = (\mathbf{T}', \tilde{\iota} \circ \alpha^{-1}) .$$

If $\alpha$ can be represented by a matrix of the form $\begin{pmatrix} \alpha_1 & 0 \\ 0 & 1 \end{pmatrix}$, then the action is given by base change with respect to the action of $\alpha_1$ on $M_I^{d-1}$. Hence we have

$$\alpha^* \mathbf{T}' = \mathbf{T}'_{(\mathbf{E}^{\mathrm{univ}}, \iota \circ \alpha_1^{-1})} ,$$

which is $\mathbf{T}'$ endowed with the level structure $\tilde{\iota}_1(x, y) = \iota \circ \alpha_1^{-1}(x) + \tilde{\iota}(y)$, where $x \in (I^{-1}/A)^{d-1}$ and $y \in I^{-1}/A$.

If $\alpha$ can be represented by a matrix of the form $\begin{pmatrix} 1_{d-1} & \alpha' \\ 0 & 1 \end{pmatrix}$, it acts relative $M_I^{d-1}$ on all spaces

$$\mathbf{E}_{\mathfrak{a}}^{\mathrm{univ}} \to \bar{\mathbf{E}}_{\mathfrak{a}}^{\mathrm{univ}} \leftarrow \mathfrak{M}_I^{d-1} .$$

Let $\nu_0 \in \mathbf{E}_{\mathfrak{a}}^{\mathrm{univ}}(M_I^{d-1})$ be the section inducing $I^{-1} \overset{\iota \circ \alpha' \circ pr}{\longrightarrow} \mathbf{E}^{\mathrm{univ}}(M_I^{d-1})$. Then $\alpha$ acts on $\mathbf{E}_{\mathfrak{a}}^{\mathrm{univ}}$ as $id - \nu_0$. Hence over suitable open subsets of $M_I^{d-1}$, the action is given by $\alpha^\sharp(X) = X - \nu_0^\sharp(X)$. Since $\nu_0$ is annihilated by $I$, we have

$$u_\Lambda\left(\frac{1}{X - \nu_0^\sharp}, Z\right) = u_\Lambda ,$$

so we have $\alpha^*\mathbf{T}' = \mathbf{T}'$. The level structure is given locally by

$$\begin{aligned}\tilde{\iota}_1(x,\bar{y}) &= u_\Lambda\left(\frac{1}{X}, \iota(x) - \varphi_y^\sharp(X - \nu_0^\sharp)\right)\\ &= \tilde{\iota}(x,\bar{y}) - u_\Lambda\left(\frac{1}{X}, \iota\circ\alpha'(\bar{y})\right)\\ &= \tilde{\iota}\circ\alpha^{-1}(x,\bar{y})\,.\end{aligned}$$

This completes the proof. □

Hence the map $\vartheta_{\mathbf{E}_a^{\mathrm{univ}}}$ induces a $\overline{\mathrm{GL}}_d^0$-equivariant map

$$\mathcal{O}(M_I^d) \xrightarrow{\vartheta} \mathcal{O}(<\mathfrak{M}_I^{d-1,1}>)\,.$$

This map corresponds to a pseudo-Drinfeld module of rank $d$ and level $I$ on $<\mathfrak{M}_I^{d-1,1}>$. We denote it by $\mathbf{T}$. Now let $O$ be a complete discrete valuation ring over $A$, and let $\xi \in \mathfrak{M}_I^{d-1,1}(O)^\circ$. From the commutative diagram

$$\begin{array}{ccc}\operatorname{Spf} O & \xrightarrow{\xi} & \mathfrak{M}_I^{d-1,1}\\ \big\downarrow & & \big\downarrow{\scriptstyle\hat{\rho}}\\ \operatorname{Spec} O & \longrightarrow & \bar{M}_I^{d-1,1}\end{array}$$

we obtain a morphism $<\operatorname{Spf} O> \xrightarrow{<\xi>} <\mathfrak{M}_I^{d-1,1}>$. Note that as a ringed space, $<\operatorname{Spf} O> \cong \operatorname{Spec} K$. With this identification, we can formulate the universal property of $\mathbf{T}$.

PROPOSITION 2.6. *Let $\xi \in \mathfrak{M}_I^{d-1,1}(O)^\circ$, and let $(\mathbf{F}, \tilde{\iota})$ be the Drinfeld module of rank d over $K$ associated to the Tate datum attached to $\xi$. Then we have $(\mathbf{F}, \tilde{\iota}) = <\xi>^* (\mathbf{T})$.*

## 3. Algebraization

Now we construct the compactification of $M_I^2$, such that the formal boundary $\mathfrak{M}_I^{1,1}$ is the completion along the complement of $M_I^2$.

The construction of the algebraization is based on a variant of M. Artin's theorem on the determination of a sheaf by formal data along a closed subspace [**Ar**], thm. 2.6. Its formulation and proof is deferred to appendix B. Here we only will need the following special case.

Let $X$ be a noetherian scheme, let $\mathcal{I} \subseteq \mathcal{O}_X$ be an invertible ideal, and let $Y := V(\mathcal{I})$ be the subscheme defined by $\mathcal{I}$. Then we have morphisms of formal schemes

$$X' = X - Y \xrightarrow{\rho} X \xleftarrow{\bar{\rho}} \mathfrak{X} = X_{/Y}$$

as considered already in chap. 4, sec. 1. Recall that we have defined

$$< X' > = < \mathfrak{X} > = (X, \rho_* \mathcal{O}_{X'} \otimes_{\mathcal{O}_X} \bar{\rho}_* \mathcal{O}_{\mathfrak{X}}) \,.$$

Similarly, for a coherent $\mathcal{O}_{X'}$-sheaf $\mathcal{F}'$ and a coherent $\mathcal{O}_{\mathfrak{X}}$-sheaf $\bar{\mathcal{F}}$ we define

$$< \mathcal{F}' > := \rho_* \mathcal{F}' \otimes_{\mathcal{O}_X} \bar{\rho}_* \mathcal{O}_{\mathfrak{X}}$$

and

$$< \bar{\mathcal{F}} > := \rho_* \mathcal{O}_{X'} \otimes_{\mathcal{O}_X} \bar{\rho}_* \bar{\mathcal{F}} \,.$$

If there is any doubt, we write $< \bar{\mathcal{F}} >_{\rho, \bar{\rho}}$.

We consider the category of all triples $(\mathcal{F}', \bar{\mathcal{F}}, \varphi)$, where $\mathcal{F}'$ is a coherent $\mathcal{O}_{X'}$-module, $\bar{\mathcal{F}}$ is a coherent $\mathcal{O}_{\mathfrak{X}}$-module and $< \mathcal{F}' > \xrightarrow{\varphi} < \bar{\mathcal{F}} >$ is an isomorphism of $< \mathcal{O}_{\mathfrak{X}} >$-modules. Morphisms of triples are defined in the obvious way.

PROPOSITION 3.1. *The category of coherent $\mathcal{O}_X$-modules is equivalent to the category of triples $(\mathcal{F}', \bar{\mathcal{F}}, \varphi)$.*

PROOF. See appendix, Theorem B.1 and Corollary B.2. □

Let us now collect the ingredients needed for our construction of the compactification. For short, we set $\mathfrak{M} = \mathfrak{M}^{1,1}$. Then we have the ringed spaces

$$\text{(1)} \qquad \mathbb{A}^1_A \xrightarrow{r} \mathbb{P}^1_A \xleftarrow{\bar{r}} \mathfrak{P}^1_A \longleftarrow < \mathfrak{P}^1_A > ,$$

where $\mathfrak{P}^1_A$ is the completion along $s_\infty$. To these spaces we will apply proposition 3.1. Let

$$\text{(2)} \qquad M^{1,1}_I \xrightarrow{\rho} \bar{M}^{1,1}_I \xleftarrow{\bar{\rho}} \mathfrak{M} \longleftarrow < \mathfrak{M} >$$

be sequences of spaces and morphisms constructed in sec. 1. These maps describe the formal boundary of our compactification ; moreover, we need the map

$$\text{(3)} \qquad \mathcal{O}(M^2_I) \xrightarrow{\vartheta} \mathcal{O}(< \mathfrak{M} >) \,,$$

coming from the universal pseudo-Drinfeld module $\mathbf{T}$, c.f. sec. 2; and the finite and flat morphism

$$\text{(4)} \qquad M^2_I \xrightarrow{j_a} \mathbb{A}^1_A$$

constructed in 4, sec. 2. We denote it by $j$, for short. Here $a \in A$ is a fixed non constant element. These maps will provide a coherent sheaf on $\mathbb{A}^1_A$ and the glueing map.

One piece is missing, which we will supply now. We construct a finite and flat morphism of formal schemes

$$\mathfrak{M} \xrightarrow{\bar{\jmath}} \mathfrak{P}^1_A \tag{5}$$

providing a coherent sheaf on $\mathfrak{P}^1_A$. To this end, we consider the universal pseudo-Drinfeld module $\mathbf{T}$ over $< \mathfrak{M} >$. It determines a section $t \in \mathcal{O}(< \mathfrak{M} >)$, see 4, sec. 2. In fact, one has

$$t = \vartheta \circ j^\sharp(T)\,.$$

Now we have the following lemma.

LEMMA 3.2. (i) *The element* $t^{-1}$ *is a section in* $\mathcal{O}(\mathfrak{M})$*;*
(ii) $t^{-1} \cdot \mathcal{O}_{\mathfrak{M}}$ *is an ideal of definition.*

PROOF. It will be sufficient to show that $(t^{-1})_{|\mathfrak{N}^1_I} \in \mathcal{O}(\mathfrak{N}^1_I)$. Let $\mathcal{I}_0 \subseteq \mathcal{O}(\mathfrak{N}^1_I)$ be the maximal ideal of definition, i.e. the radical of any ideal of definition. Then $V(\mathcal{I}_0) = M^1_I$ and $\mathbf{T}_{|M^1_I} = \mathbf{E}^{\mathrm{univ}}$ (by abuse of language). Taking over the notations of 4, sec. 2, we see that $a_1$ is a nowhere vanishing section and $a_2$ vanishes on $V(\mathcal{I}_0)$. Hence

$$t^{-1} = \frac{a_2^{N_2}}{a_1^{N_1}} \in \mathcal{O}(\mathfrak{N}^1_I)\,.$$

Let $\mathfrak{U} \subseteq \mathfrak{N}^1_I$ be open such that $\mathfrak{U} \cong \operatorname{Spf} R[[1/X]]$. By 2.3, we have $t = X^n \cdot b$ for some $n \in \mathbb{N}$ and $b \in \mathcal{O}(\mathfrak{U})^\times$. Hence $X^{-n} = t^{-1} \cdot b$, so $V(t^{-1}) = V(\mathcal{I}_0)$. This proves the lemma. □

Hence $t^{-1}$ defines a morphism

$$\mathfrak{M} \xrightarrow{\bar{\jmath}} \mathfrak{P}^1_A = \operatorname{Spf} A \left[\left[\frac{1}{T}\right]\right]\,.$$

Note that the topology on $\mathcal{O}(\mathfrak{M})$ coincides with the $(t^{-1})$-adic topology.

LEMMA 3.3. *The morphism* $\bar{\jmath}$ *is finite and flat.*

PROOF. First we note that $\bar{\jmath}$ is a morphism of affine formal schemes. Moreover, it will be sufficient to treat its restriction to $\mathfrak{N}^1_I$. Since $V(t^{-1}) = V(\mathcal{I}_0)$, the reduction of $(\mathfrak{N}^1_I, \mathcal{O}_{\mathfrak{N}^1_I}/(t^{-1}))$ is $M^1_I$, which is finite over $\operatorname{Spec} A$, by 3, 4.1, so the former scheme is finite, as well. Since $\bar{\jmath}_*\mathcal{O}_{\mathfrak{N}^1_I}$ is a complete module over $\mathcal{O}_{\mathfrak{P}^1_A}$, c.f. 3.2, (ii), it is finite. Locally,

$\mathcal{O}(\mathfrak{U})/(t^{-1})$ is flat over $A$, and multiplication by $t^{-1}$ is an injective map. By the local flatness criterion, see [**Mat**], 22.3, $\bar{\jmath}$ is flat. $\square$

Now we wish to apply 3.1 to the sheaves

$$\mathcal{F}' = j_*\mathcal{O}_{M_I^2} \text{ and } \bar{\mathcal{F}} = \bar{\jmath}_*\mathcal{O}_{\mathfrak{M}}\,.$$

The problem is that we have to construct an isomorphism

$$\varphi : <\mathcal{F}'>_{r,\bar{r}} \longrightarrow <\hat{\mathcal{F}}>_{r,\bar{r}}\ .$$

First, the map $\vartheta$, (3), induces a map $\mathcal{O}(M_I^2)\otimes_A\mathcal{O}_{<\mathfrak{P}^1>} \xrightarrow{\varphi''} \bar{\jmath}_*\mathcal{O}_{<\mathfrak{M}>_{\rho,\bar{\rho}}}$, which factorizes through $\mathcal{O}(M_I^2)\otimes_{A[T]}\mathcal{O}_{<\mathfrak{P}^1>} \xrightarrow{\varphi'} \bar{\jmath}_*\mathcal{O}_{<f\mathcal{M}>_{\rho,\bar{\rho}}}$. Now there is a canonical isomorphism

$$\mathcal{O}(M_I^2)\otimes_{A[T]}\mathcal{O}_{<\mathfrak{P}^1>} \cong < j_*\mathcal{O}_{M_I^2}>_{r,\bar{r}}\ .$$

In fact, since $\mathcal{O}(M_I^2)$ is flat over $A[T]$, there is an isomorphism $j_*\mathcal{O}_{M_I^2} \cong \mathcal{O}(M_I^2)\otimes_{A[T]}\mathcal{O}_{\mathbb{A}^1_A}$. This implies the result. Composing with this isomorphism, $\varphi'$ gives rise to a map of $\mathcal{O}_{<\mathfrak{P}^1_A>}$-algebras

$$< j_*\mathcal{O}_{M_I^2}>_{r,\bar{r}} \xrightarrow{\ \varphi\ } <\bar{\jmath}_*\mathcal{O}_{\mathfrak{M}}>_{r,\bar{r}}\ .$$

Now we have to check that $\varphi$ is an isomorphism. Let $x \in \operatorname{Spec} A$ be any point not contained in $V(I)$. Then $C = M_I^2 \times_A \kappa(x)$ is a smooth curve by 3, 4.1, (iii). It has a unique smooth compactification $\bar{C}$. The map $j$ extends as a map into $\mathbb{P}^1_{\kappa(x)}$, and the corresponding formal boundary identifies canonically with $\mathfrak{M}\times_A\kappa(x)$, see 1.11. The situation is visualized in the commutative diagram

$$\begin{array}{ccccccc}
C & \longrightarrow & \bar{C} & \longleftarrow & \hat{C} & \longleftarrow & \mathfrak{M}\times_A\kappa(x) \\
\big\downarrow{\scriptstyle j} & & \big\downarrow{\scriptstyle \bar{\jmath}} & & \big\downarrow{\scriptstyle \bar{\jmath}'} & \swarrow{\scriptstyle \bar{\jmath}} & \\
\mathbb{A}^1_{\kappa(x)} & \xrightarrow{\ r\ } & \mathbb{P}^1_{\kappa(x)} & \xleftarrow{\ \bar{r}\ } & \mathfrak{P}^1_{\kappa(x)} & &
\end{array}$$

Hence the triple $(j_*\mathcal{O}_{M_I^2\times\kappa(x)}, \bar{\jmath}_*\mathcal{O}_{\mathfrak{M}\times\kappa(x)}, \varphi\times\kappa(x))$ identifies with $\Phi(\bar{\jmath}_*\mathcal{O}_{\bar{C}})$, c.f. 3.1, where $\Phi$ is the functor from appendix B, theorem B.1. We conclude that $< j_*\mathcal{O}_{M_I^2}>$ and $<\bar{\jmath}_*\mathcal{O}_{\mathfrak{M}}>$ are locally free $\mathcal{O}_{<\mathfrak{P}^1_A>}$-sheaves of the same rank, and $\varphi$ is injective.

Let $y$ be any point of $\mathfrak{P}^1_A = \operatorname{Spf} A[1/T]$, and let $L$ be the quotient field of $B = \mathcal{O}_{<\mathfrak{P}^1_A>,y}$. Then $\varphi_y\otimes_B L$ is bijective, being an injective homomorphism of $L$-vector spaces of the same dimension. In order to derive bijectivity of $\varphi_y$ from this, we prove the following lemma.

LEMMA 3.4. *Both $B$-algebras, $< j_*\mathcal{O}_{M_I^2}>_y$ and $<\bar{\jmath}_*\mathcal{O}_{\mathfrak{M}}>_y$, are finite products of normal integral domains.*

PROOF FOR $< j_*\mathcal{O}_{M_I^2} >$. We have

$$< j_*\mathcal{O}_{M_I^2} >_y = \varinjlim_{s \notin y} \mathcal{O}(M_I^2) \otimes_{A[T]} A_s\left(\left(\frac{1}{T}\right)\right).$$

But by regularity, 3, 4.1, $\mathcal{O}(M_I^2)$ is a product of normal integral domains. Let $C$ be one of its factors. Then we are going to show that $C \otimes_{A[T]} A_s((1/T))$ is a normal integral domain. Let $D$ be the integral closure of $A[1/T]$ in $Q(C)$. Since $A[1/T]$ is excellent, [**Mat1**], chap. 13, 33.H, $D$ is finite. In particular it is excellent. Now we have

$$C \otimes_{A[T]} A_s\left(\left(\frac{1}{T}\right)\right) = D \otimes_{A[1/T]} A_s\left(\left(\frac{1}{T}\right)\right).$$

Hence it is the localization by $1/T$ of the completion of $D_s$ with respect to $(1/T)$. This is a normal integral domain, loc. cit. 33.A.

*Proof for* $< \bar{\jmath}_*\mathcal{O}_{\mathfrak{M}} >$. Let $z \in V(I)$ be different from $y$, and let $\mathfrak{U} = \mathfrak{P}_A^1 - \{z\}$. Then we have

$$< \bar{\jmath}_*\mathcal{O}_{\mathfrak{M}} >_y = \varinjlim_{s \notin y} (\mathcal{O}_{\mathfrak{M}}(\bar{\jmath}^{-1}(\mathfrak{U})) \hat{\otimes}_A A_s)_{1/T}.$$

But thanks to regularity of $M_I^1$, $\mathcal{O}_{\mathfrak{M}}(\bar{\jmath}^{-1}(\mathfrak{U}))$ is a product of algebras of the form $R[[1/X]]$, where $R$ is a normal integral domain. This proves the result. □

Now we can finish to prove that $\varphi_y$ is bijective. Since $\varphi \otimes L$ is bijective, $\varphi_y$ is a product of homomorphisms of normal domains, which become isomorphisms after tensoring with $L$. It follows that they are isomorphisms.

Now we put $\mathcal{B} := \Psi(j_*\mathcal{O}_{M_I^2}, \bar{\jmath}_*\mathcal{O}_{\mathfrak{M}}, \varphi)$, 3.1, where again $\Psi$ is the functor from appendix B, theorem B.1. It is a coherent $\mathcal{O}_{\mathbb{P}_A^1}$-algebra. Then we have constructed a compactification of $M_I^2$ setting

$$\bar{M}_I^2 := \mathbf{S}pec\ \mathcal{B}.$$

We summarize the results of this chapter. Recall that $r$ is the degree of the residue field $\kappa(\infty)$ over $\mathbb{F}_q$ and $k = Q(A)$.

PROPOSITION 3.5. *$\bar{M}_I^2$ is a regular scheme, which is finite over $\mathbb{P}_A^1$. In particular, it is proper over* Spec $A$. *It contains $M_I^2$ as an open and dense subspace, and the complement $\bar{M}_I^2 - M_I^2$ is a disjoint union of schemes, which are isomorphic to $M_I^1$. More precisely, there are*

$$n_I = \frac{|\mathrm{GL}_2(A/I)|}{|(A/I)^\times| \cdot |A/I|} \cdot \frac{r \cdot \mathrm{Cl}(k)}{q-1}$$

*copies of $M_I^1$. Finally, $\bar{M}_I^2$ is smooth over* Spec $A - V(I)$.

PROOF. By 3, 4.1, $\bar{M}_I^2$ is regular at all points of $M_I^2$. Let $x$ be a closed point of $\bar{M}_I^2 - M_I^2$. Then $x$ can be considered as a closed point of $\mathfrak{M} = \mathfrak{M}_I^2$. Since $\mathcal{O}_{\mathfrak{M},x} \cong \mathcal{O}_{M_I^1,y}[[T]]$ for some closed point $y$, we conclude that $\hat{\mathcal{O}}_{M_I^2,x} \cong \hat{\mathcal{O}}_{M_I^1,y}[[T]]$ is regular. This proves regularity of $\bar{M}_I^2$. The formula for $n_I$ follows from the proofs of 1.9 and 1.8. Smoothness of $M_I^2$ and $M_I^1$ over $\operatorname{Spec} A - V(I)$ imply the last assertion. $\square$

Now, a posteriori, we can give an elementary construction of $\bar{M}_I^2$. Let $\mathcal{O}(M_I^2) = \prod C_i$ be the decomposition into its irreducible components, and let $D_i$ be the integral closure of $A[1/T]$ in $Q(C_i)$. Then the coherent sheaves $\prod \tilde{C}_i$ on $\operatorname{Spec} A[T]$ and $\prod \tilde{D}_i$ on $\operatorname{Spec} A[1/T]$ glue to a coherent sheaf $\mathcal{B}$ on $\mathbb{P}_A^1$, and we have $\bar{M}_I^2 = \mathbf{S}pec\ \mathcal{B}$. Note that this construction neither shows regularity of $\bar{M}_I^2$ nor gives a modular description of the boundary. Nevertheless, it extends to arbitrary rank, using the finite map $M_I^d \xrightarrow{j_a} \mathbb{A}_A^{d-1}$, 4, 2.3. Let $\mathbb{P}_A^{d-1} = \bigcup U_j$ be the standard affine covering, and let $\mathcal{O}(M_I^d) = \prod C_i$ be the decomposition into its irreducible components. Then the $C_i$ are normal integral domains. Let $D_{ij}$ be the integral closure of $\mathcal{O}(U_j)$ in $Q(C_i)$. The coherent sheaves $\prod_i \tilde{D}_{ij}$ on $U_i$ glue to a coherent algebra $\mathcal{B}$ on $\mathbb{P}_A^{d-1}$, and $M_I^{d-1}$ is embedded as an open and dense subspace in $\bar{M}_I^{d-1} := \mathbf{S}pec\ \mathcal{B}$, which is proper over $\operatorname{Spec} A$. By the same method, one can even embed $M_I^{d-1}$ into a space, which is proper over the curve $\mathcal{C}$.

Using this construction, it is trivial to see that the action of $\mathrm{GL}_2(\hat{A})$ extends to an action on $\bar{M}_I^2$. In fact, it acts relative to $\operatorname{Spec} A[T]$ on the moduli space. Thus it acts on the integral closure of $A[1/T]$ in $\prod Q(C_i)$, as well. This implies the result. Similarly, it follows that for $J \subseteq I$, the restriction map $M_J^2 \to M_I^2$ extends to $\bar{M}_J^2 \to \bar{M}_I^2$, which is finite and flat (flatness follows from regularity). Moreover, one has $\bar{M}_I^2 = \Gamma(I)\backslash \bar{M}_J^2$. Finally, we have an embedding

$$M^2 \longrightarrow \bar{M}^2 := \varprojlim_I \bar{M}_I^2 \,,$$

which induces an isomorphism

$$\bar{M}_I^2 \cong \Gamma(I)\backslash M^2 \,.$$

# Appendix

## A. Induced schemes

Let $G$ be a group, let $H$ be a subgroup of $G$, which acts on a scheme $X$. Then there is a scheme $G \times^H X$, endowed with a $G$-action, and a $H$-equivariant map $\rho : X \to G \times^H X$ having the following universal property:

Given a scheme $Y$, a $G$-action on $Y$ and a $H$-equivariant map $\varphi : X \to Y$, then there is a unique $G$-equivariant map $\psi : G \times^H X \to Y$ such that $\psi \circ \rho = \varphi$.

In fact, let $[\bar{q}] \in G$ be a fixed representative of a coset $\bar{q} \in G/H$. We assume that $[\bar{1}] = 1$. Then we define

$$G \times^H X = \dot{\bigcup_{\bar{q} \in G/H}} X_{\bar{q}},$$

where $X_{\bar{q}} = X$ (disjoint union). Let $\sigma_{\bar{q}} : X = X_{\bar{1}} \to X_{\bar{q}}$ be the identity. Then we define an action of $G$ on that scheme by

$$g_{|X_{\bar{q}}} = \sigma_{\bar{g}\bar{q}} \circ ([\bar{g}\bar{q}]^{-1} g [\bar{q}]) \circ \sigma_{\bar{q}}^{-1}.$$

It is immediate to verify that this gives a $G$-action, and that this scheme together with the embedding $\sigma_{\bar{1}}$ has the desired property.

If $H$ acts on a formal scheme $\mathfrak{X}$, the same construction applies. For any open subset $\mathfrak{U}$ of $\mathfrak{X}$, the ring $\mathcal{O}_{\mathfrak{X}}(\mathfrak{U})$ carries the coarsest topology, such that the restriction maps $\mathcal{O}(\mathfrak{U}) \to \mathcal{O}(\mathfrak{X}_{\bar{q}} \cap \mathfrak{U})$ are continuous.

For any scheme $T$, there is a canonical bijection of $G$-sets

$$(G \times^H X)(T) \cong G \times^H X(T)$$

and similar for formal schemes.

Let $K \subseteq G$ be another subgroup, and suppose that for all $g \in G$ the categorical quotient $(H \cap g^{-1}Kg)\backslash X$ exists. Then the categorical quotient $K\backslash(G \times^H X)$ exists, and we have

$$K\backslash(G \times^H X) \cong \bigcup_{Q \in K\backslash G/H} (H \cap q^{-1}Kq)\backslash X_{\bar{q}}.$$

Here $q$ is a fixed representative of $Q$. The same holds for geometrical quotients. In particular for a normal subgroup $K$ one has

$$K\backslash(G \times^{H} X) = (G/K) \times^{H/K\cap H} (K \cap H\backslash X)\,.$$

## B. Construction of coherent sheaves

Here we prove a variant of M. Artin's theorem on the determination of a sheaf by formal data along a closed subspace [**Ar**], thm. 2.6. In our form, it holds for arbitrary noetherian schemes.

Let $X$ be a noetherian scheme, and let $Y \subseteq X$ be a closed subscheme. We denote by

$$\begin{array}{ll} X' : & \text{the complement } X\backslash Y; \\ \mathcal{I} : & \text{the ideal of } Y; \\ \mathfrak{X} : & \text{the completion } X_{/Y}. \end{array}$$

There are canonical mappings

$$X' = X - Y \xrightarrow{\rho} X \xleftarrow{\bar{\rho}} \mathfrak{X}\,.$$

Recall that $< X' >$ is the ringed space $(X, \rho_* \mathcal{O}_{X'} \otimes_{\mathcal{O}_X} \bar{\rho}_* \mathcal{O}_{\mathfrak{X}})$.

GENERAL REMARK. Let $\mathcal{N}$ be a quasi-coherent $\mathcal{O}_X$-module with supports contained in $Y$. Then $\mathcal{N} \otimes_{\mathcal{O}_X} \bar{\rho}_* \mathcal{O}_{\mathfrak{X}} \cong \mathcal{N}$. Thus a coherent $\mathcal{O}_X$-module with supports in $Y$ is the same as a coherent $\mathcal{O}_{\mathfrak{X}}$-module which is annihilated by some power of $\mathcal{I}\mathcal{O}_{\mathfrak{X}}$.

Now we consider the following category:

Objects are triples $(\mathcal{F}', \bar{\mathcal{F}}, \varphi)$, where

- $\mathcal{F}'$ is a coherent $\mathcal{O}_{X'}$-module,
- $\bar{\mathcal{F}}$ is a coherent $\mathcal{O}_{\mathfrak{X}}$-module,
- $\varphi$ is a homomorphism from $\bar{\rho}_* \bar{\mathcal{F}}$ to $< \mathcal{F}' > = \rho_* \mathcal{F}' \otimes_{\mathcal{O}_X} \bar{\rho}_* \mathcal{O}_{\mathfrak{X}}$ such that both $\ker(\varphi)$ and $\operatorname{coker}(\varphi)$ are quasi-coherent $\mathcal{O}_X$-modules with supports in $Y$.

Morphisms are pairs $\mathcal{F}' \xrightarrow{\alpha'} \mathcal{G}'$, $\bar{\mathcal{F}} \xrightarrow{\bar{\alpha}} \bar{\mathcal{G}}$ making the diagram

$$\begin{array}{ccc} \bar{\rho}_* \bar{\mathcal{F}} & \xrightarrow{\bar{\rho}_* \bar{\alpha}} & \bar{\rho}_* \bar{\mathcal{G}} \\ \downarrow \varphi & & \downarrow \psi \\ < \mathcal{F}' > & \xrightarrow{<\alpha'>} & < \mathcal{F}' > \end{array}$$

commute.

THEOREM B.1. *The category of coherent* $\mathcal{O}_X$*-modules is equivalent to the category of triples* $(\mathcal{F}', \bar{\mathcal{F}}, \varphi)$.

PROOF. Let $\mathcal{F}$ be a coherent $\mathcal{O}_X$-module. Then we define a triple $(\mathcal{F}', \bar{\mathcal{F}}, \varphi_{\mathcal{F}})$ by $\mathcal{F}' := \rho^*\mathcal{F}$, and $\bar{\mathcal{F}} := \bar{\rho}^*\mathcal{F}$. Note that $\bar{\rho}_*\bar{\mathcal{F}} \cong \mathcal{F} \otimes_{\mathcal{O}_X} \bar{\rho}_*\bar{\mathcal{O}}_{\mathfrak{X}}$. Let $\mathcal{F}' \xrightarrow{\varphi'} \rho_*\mathcal{F}$ be the canonical homomorphism. Then $\varphi_{\mathcal{F}} := \varphi' \otimes id_{\bar{\rho}_*\mathcal{O}_{\mathfrak{X}}}$. Since both kernel and cokernel of $\varphi'$ are quasi-coherent $\mathcal{O}_X$-modules with supports in $Y$, $(\mathcal{F}', \bar{\mathcal{F}}, \varphi_{\mathcal{F}})$ is really a triple. Clearly this gives a functor $\Phi$ from coherent $\mathcal{O}_X$-modules to the category of triples. Denote $\mathcal{K} := \ker(\varphi')$, $\mathcal{Q} := \operatorname{coker}(\varphi')$. Then one obtains the following commutative diagram, where one has to take into acccunt additionally, that $\bar{\rho}_*\mathcal{O}_{\mathfrak{X}}$ is flat over the noetherian algebra $\mathcal{O}_X$, and the Remark above, as $\mathcal{K}$, $\mathcal{Q}$ have support in $Y$.

$$\begin{array}{ccccccccc}
0 & \longrightarrow & \mathcal{K} & \longrightarrow & \mathcal{F} & \xrightarrow{\varphi'} & \rho_*\mathcal{F}' & \longrightarrow & \mathcal{Q} & \longrightarrow & 0 \\
 & & \| & & \downarrow & & \downarrow & & \| & & \\
0 & \longrightarrow & \mathcal{K} & \longrightarrow & \bar{\rho}_*\bar{\mathcal{F}} & \xrightarrow{\varphi_{\mathcal{F}}} & < \mathcal{F}' > & \longrightarrow & \mathcal{Q} & \longrightarrow & 0
\end{array}.$$

Now we will construct a quasi-inverse of $\Phi$. We proceed in several steps.

*Step 1.* Let $(\mathcal{F}', \bar{\mathcal{F}}, \varphi)$ be a triple. Then we define

$$\Psi(\mathcal{F}', \bar{\mathcal{F}}, \varphi) := \mathcal{F} := \ker\left(\rho_*\mathcal{F}' \oplus \bar{\rho}_*\bar{\mathcal{F}} \xrightarrow{\delta} < \mathcal{F}' >\right),$$

where $\delta(f, g) = f \otimes 1 - \varphi(1 \otimes g)$. Clearly $\Psi$ gives a functor from the category of triples to the category of $\mathcal{O}_X$-modules. We will show that $\mathcal{F}$ is actually coherent and that $\Psi$ is quasi-inverse to $\Phi$.

*Step 2.* Let $\mathcal{F}$ be a coherent $\mathcal{O}_X$-module. Then $\Psi(\Phi(\mathcal{F})) \cong \mathcal{F}$. More precisely, there is an exact sequence

$$0 \to \mathcal{F} \to \rho_*\mathcal{F}' \oplus \bar{\rho}_*\bar{\mathcal{F}} \xrightarrow{\delta} < \mathcal{F}' > \to 0.$$

In fact, it follows from the above diagram of exact sequences that the square in the middle is cartesian and cocartesian. This is equivalent to this sequence being exact.

*Step 3.* There exists a coherent $\mathcal{O}_X$-module $\mathcal{F}_0$ and a morphism $\Phi(\mathcal{F}_0) \xrightarrow{(\alpha', \bar{\alpha})} (\mathcal{F}', \bar{\mathcal{F}}, \varphi)$, such that $\alpha'$ is an isomorphism.

To construct such a module, we choose any coherent submodule $\mathcal{F}_0 \subseteq \rho_*\mathcal{F}'$ with $(\mathcal{F}_0)_{|X'} = \mathcal{F}'$, [**Ha**], chap. II, exerc. 5.15. Consider the undotted arrows of the diagram

$$\begin{array}{ccccccccc}
0 & \longrightarrow & \mathcal{K}_0 & \longrightarrow & \bar{\rho}_*\bar{\mathcal{F}}_0 & \xrightarrow{\varphi_{\mathcal{F}_0}} & < \mathcal{F}_0' > & \longrightarrow & \mathcal{Q}_0 & \longrightarrow & 0 \\
 & & \vdots\,\gamma & & \vdots\,\bar{\alpha} & & \| & & \vdots\,\beta & & \\
0 & \longrightarrow & \mathcal{K} & \longrightarrow & \bar{\rho}_*\bar{\mathcal{F}} & \xrightarrow{\varphi} & < \mathcal{F}' > & \longrightarrow & \mathcal{Q} & \longrightarrow & 0
\end{array}.$$

The image of of $\bar{\rho}_*\bar{\mathcal{F}}_0$ in $\mathcal{Q}$ is annihilated by some power of $\mathcal{I}$. Replacing $\mathcal{F}_0$ by a suitable $\mathcal{I}^N\mathcal{F}_0$, we may assume that there is an induced homomorphism $\mathcal{Q}_0 \xrightarrow{\beta} \mathcal{Q}$. By the Artin-Rees lemma there exists a number $M$ such that $(\mathcal{I}^M\bar{\rho}_*\bar{\mathcal{F}}) \cap \mathcal{K} = (0)$. Replacing again $\mathcal{F}_0$ by $\mathcal{I}^M\mathcal{F}_0$, the diagram can be completed by homomorphisms $\bar{\alpha}$ and $\gamma = 0$. This gives the desired morphism $(\mathcal{F}_0', \bar{\mathcal{F}}_0, \varphi_{\mathcal{F}_0}) \to (\mathcal{F}', \bar{\mathcal{F}}, \varphi)$.

*Step 4.* By construction, ker $\bar{\alpha} \cong \mathcal{K}_0$. Hence it is coherent with supports contained in $Y$. On the other hand, coker $\bar{\alpha}$ sits in an exact sequence

$$0 \to \mathcal{K} \to \text{coker } \bar{\alpha} \to \ker \beta \to 0\,.$$

Hence coker $\bar{\alpha}$ is quasi-coherent with supports in $Y$. But it is finitely generated as $\bar{\rho}_*\mathcal{O}_{\mathfrak{X}}$-module. Therefore it is even coherent. The morphism $(\alpha', \bar{\alpha})$ of Step 3 induces the commutative diagram below, where $\mathcal{F} = \Psi(\mathcal{F}', \bar{\mathcal{F}}, \varphi)$:

$$\begin{array}{ccccccccc}
0 & \longrightarrow & \mathcal{F}_0 & \longrightarrow & \rho_*\mathcal{F}_0' \oplus \bar{\rho}_*\bar{\mathcal{F}}_0 & \xrightarrow{\delta_0} & <\mathcal{F}_0'> & \longrightarrow & 0 \\
 & & \downarrow{\scriptstyle \alpha} & & \downarrow{\scriptstyle \alpha' \oplus \bar{\alpha}} & & \downarrow{\scriptstyle <\alpha>} & & \\
0 & \longrightarrow & \mathcal{F} & \longrightarrow & \rho_*\mathcal{F}' \oplus \bar{\rho}_*\bar{\mathcal{F}} & \xrightarrow{\delta} & <\mathcal{F}'> & &
\end{array}\,.$$

Since $\alpha'$ is an isomorphism, we see that $\delta$ is surjective and that ker $\alpha \cong$ ker $\bar{\alpha}$ and coker $\alpha \cong$ coker $\bar{\alpha}$, both being coherent. Hence $\mathcal{F}$ is coherent as well. In particular, starting now again with $\mathcal{F}_0$ above replaced by the coherent $\mathcal{F} = \Psi(\mathcal{F}', \bar{\mathcal{F}}, \varphi)$ just constructed, we obtain the corresponding diagram above in this new situation. To be able to do this, one has to observe, that $\mathcal{F} \cong \bar{\rho}^*\mathcal{F}$. To see this one considers the $\mathcal{I}$-adic completion of the second row in the diagram above. From this follows the isomorphism $\Phi(\mathcal{F}) \cong (\mathcal{F}', \bar{\mathcal{F}}, \varphi)$ which we wanted. This completes the proof. □

COROLLARY B.2. *Let $\mathcal{I} = \mathcal{I}(Y)$ be locally a principal ideal. Then the category of coherent $\mathcal{O}_X$-modules is equivalent to the category of triples $(\mathcal{F}', \bar{\mathcal{F}}, \varphi)$, where $\mathcal{F}'$ and $\bar{\mathcal{F}}$ are as above and $<\bar{\mathcal{F}}> \xrightarrow{\varphi} <\mathcal{F}'>$ is an isomorphism of $<\mathcal{O}_{\mathfrak{X}}>$-modules.*

PROOF. In this case $\rho_*\mathcal{O}_{X'}$ is a flat $\mathcal{O}_X$-module and $\mathcal{N} \otimes_{\mathcal{O}_X} \rho_*\mathcal{O}_{X'} = 0$ for any quasi-coherent $\mathcal{O}_X$-module $\mathcal{N}$ with supports in $Y$. Hence any morphism $\bar{\rho}_*\bar{\mathcal{F}} \xrightarrow{\varphi} <\mathcal{F}'>$ as in the theorem induces an isomorphism $<\bar{\mathcal{F}}> \xrightarrow{\varphi} <\mathcal{F}'>$ and vice versa. □

# Bibliography


[Ar] Artin, M. *Algebraization of formal moduli II,* Ann. Math. 91, (1970), 88-135.

[BGR] Bosch, S.; Güntzer; U., Remmert, R. *Non-Archimedean Analysis,* Grundlehren der Math. Wiss. 261, Springer-Verlag Berlin, (1984).

[Bou] Bourbaki, N. *Commutative Algebra,* Springer-Verlag, Berlin, (1989).

[Dr] Drinfeld, V.G. *Elliptic Modules,* Math. USSR Sbornic 23, (1976), 561-592.

[Du] Dummit, D.S. *Genus Two Hyperelliptic Drinfeld Modules over* $\mathbb{F}_2$, in [**GHR**], 117-129.

[EGA] Grothendieck, A. *Elements de geometrie algebrique I-IV,* Publ. IHES, 4, 8, 11, 17, 20, 24, 28, (1960-1967).

[Ge] Gekeler, E.U. *On the de Rham isomorphism for Drinfeld modules,* J. reine angew. Math. 401, (1989), 188-208.

[Ge1] Gekeler, E.U. *Drinfeld modular curves,* Lecture Notes in Math. 1231, Springer-Verlag, Berlin, (1986).

[GPRV] Gekeler, E.U.; van der Put; M. Reversat, M.; Van der Geel, J. (Ed.) *Drinfeld Modules, Modular Schemes and Applications,* World Scientific Publ., (1997).

[Go] Goss, D. *Basic Structures of Function Field Arithmetic,* Springer-Verlag Berlin, (1996).

[GHR] Goss, D.; Hayes, D.R.; Rosen, M.I. *The Arithmetic of Function Fields,* Ohio State Univ., Math Res. Inst. Publ. 2, de Gruyter, (1992).

[Ha] Hartshorne, R. *Algebraic Geometry* Graduate Texts in Math. 11, Springer-Verlag Berlin, (1977).

[Hasse] Hasse, H. *Zahlentheorie,* Akademie-Verlag, Berlin, (1969).

[Ka] Kapranov, M.M. *On Cuspidal Divisors on the Modular Varieties of Elliptic Modules,* Math. USSR Izvestiya 30, (1988), 533-547.

[KLS] Kumar, S.; Laumon, G.; Stuhler, U. *Vector Bundles on Curves- New Directions, Cetraro 1995,* Lecture Notes in Math. 1649, Springer-Verlag, Berlin, (1997).

[La] Laumon, G. *Cohomology of Drinfeld Modular Varieties,* Part I, Cambridge stud. in adv. math. 41, (1996).

[Le] Lehmkuhl, T. *Compactification of the Drinfeld modular surfaces,* Habilitationsschrift, Göttingen, (2000).

[Mat] Matsumura, H. *Commutative Ring Theory,* Cambridge Univ. Press, (1986).

[Mat1] – *Commutative algebra,* Benjamin, (1970).

[MFK] Mumford, D.; Fogarty, J.; Kirwan, F. *Geometric Invariant Theory,* Ergebnisse der Math. 34, 3. Edition, Springer-Verlag, Berlin, (1994).

[S] Schlessinger, M. *Functors of Artin Rings,* Trans. AMS 130, (1968), 208-222.

[Se] Serre, J.P. *Algébre locale,* Lecture Notes in Math. 11, Springer-Verlag, Heidelberg, (1965).

[vHei] van der Heiden, G.-J. *Drinfeld modular curve and Weil pairing,* arXiv: math. AG/0411490 v1, 22.11.2004

[Weil] Weil, A. *Basic Number Theory,* Grundlehren der math. Wiss. 144, Springer-Verlag, Heidelberg, (1967).

# Glossary of Notations

| | | |
|---|---|---|
| $f^\sharp$ | morphism of the structure sheaves, | 1 |
| $\mathbb{A}^1_S$ | the affine line over $S$, | 1 |
| $\lvert M \rvert$ | cardinality of a set, | 1 |
| $\mathbb{G}_{a,\mathcal{L}}$, $\mathbb{G}_{a,R}$ | line bundle, | 1 |
| $\mathrm{Frob}_S$ | Frobenius, | 2 |
| $\tau_X$ | relative Frobenius, | 2 |
| $\tau_{X,q} = \tau_q$, $q = p^m$ | $\mathbb{F}_q$-linear Frobenius, | 2 |
| $R\{\tau_q\}$ | ring of $\mathbb{F}_q$-linear endomorphisms of $\mathbb{G}_{a,R}$, | 4 |
| $\partial$ | initial term of an endomorphism, | 5 |
| $D\varphi^\sharp$ | derivative of $\varphi^\sharp$, | 5 |
| $T_{1,K} = T_1$ | Tate algebra, | 14 |
| $\overset{\circ}{T}_1$ | the lattice in $T_1$, | 14 |
| $\infty$ | the distinguished point on a curve, | 15 |
| $\lvert \cdot \rvert_{\mathfrak{p}}$ | the normalized norm to a closed point $\mathfrak{p}$ on a curve, | 15 |
| $u_\Lambda$ | the exponential series to a lattice, | 16 |
| $\mathbf{E}$, $e_a$ | Drinfeld module, | 19 |
| $\mathbf{E}(T)$, $\mathbf{E}(B)$ | the module of $T$-valued points, | 20 |
| $\mathbf{E}(\bar{\ell})_{\mathrm{tors}}$ | the torsion submodule, | 21 |
| $\mathbf{E}(\bar{\ell})_{\mathfrak{q}-\mathrm{tors}}$ | the $\mathfrak{q}$-torsion module, | 21 |
| $\mathrm{rk}(\mathbf{E})$ | the rank of a Drinfeld module, | 21 |
| $\mathbf{E}[I]$ | the subscheme of $I$-division points, | 25 |
| $\mathbb{M}^d_I$, $M^d_I$ | the functor "Drinfeld modules of rank $d$ and level $I$" and its moduli space, | 28 |
| $M^d$ | the projective limit of all $M^d_I$, | 29 |
| $\mathrm{End}^{\mathrm{bdl}}(\mathrm{S}_{\mathcal{O}_X}(\mathcal{L}^{-1}))$ | the additive endomorphisms, | 31 |
| $\mathcal{C}_O$, $\hat{\mathcal{C}}_O$ | a category of artinian, complete $O$-algebras, respectively, | 33 |
| $\mathrm{Def}_{\mathbf{E}_0}$ | the functor "deformations of $\mathbf{E}_0$", | 33 |
| $\mathfrak{t}_{\mathbf{E}_0}$, $\mathfrak{t}^+_{\mathbf{E}_0}$ | | 34 |
| $\mathbf{E}/I$ | $\mathbf{E}/\mathbf{E}[I]$, | 36 |

| | |
|---|---|
| $\mathbb{C}_\infty$ | 38 |
| $\tilde{\mathfrak{t}}^+_{add}$ | 39 |
| $\mathrm{Def}_{\varphi_0}$ | 40 |
| $t_{\mathbf{E}_0}$, $t_{\varphi_0}$ | 41 |
| $d\varphi_0$, $\delta(\varphi_0)$ | 41 |
| $\mathrm{Def}_{\mathbf{E}_0,\iota_0}$ | 42 |
| $\mathbb{A}_{\mathrm{f}}$ | the ring of finite adeles, 47 |
| $\Gamma(I)$, $\Gamma_{I,J}$ | the congruence subgroup, 49 |
| $\mathrm{GL}^0_d$ | $\mathrm{GL}_d(\hat{A})\cdot\mathbb{A}_{\mathrm{f}}^\times$, 50 |
| $\mathrm{Spf}\,(R,I)=\mathrm{Spf}\,R$ | the formal spectrum of $R$, 51 |
| $R_{\{f\}}$, $R_{\{\mathfrak{p}\}}$ | 51 |
| $X_{/Y}$ | the completion of $X$ along $Y$, 52 |
| $\mathfrak{X}(O)^\circ$ | the $\mathrm{Spf}\,O$-valued points, which do not extend to $\mathrm{Spec}\,O$, 53 |
| $<\mathfrak{X}>$, $<X'>$ | a structure of ringed space on $\mathfrak{X}$, 53 |
| $j_a$ | 56 |
| $(\mathbf{E},\,\Lambda,\,\nu)$ | normal Tate datum, 58 |
| $(\mathbf{E},\,\iota;\,\Lambda,\,\nu,\,\kappa)$ | Tate datum of level $I$, 58 |
| $u_\Lambda$ | the exponential series to a lattice, 59 |
| $D^\circ$ | the integral part of a divisor, 61 |
| $u^*(D)$, $u_*(D)$ | the pullback and push down of an integral divisor, 61 |
| $\overline{\mathrm{GL}}^0_d$ | 64 |
| $\bar{G}_{d_1}$ | 66 |
| $\mathrm{NT}^{d_1,d_2}_I(O)$ | normalized Tate data of rank $d_1, d_2$ and level $I$, 66 |
| $\mathrm{Tate}^{d_1,d_2}_I(O)$ | Tate data of rank $d_1, d_2$ and level $I$, 66 |
| $\mathbf{E}^{\mathrm{univ}}_{\mathfrak{a}}$ | $\mathbf{E}^{\mathrm{univ}}/\mathfrak{a}$, 67 |
| $\bar{\mathbf{E}}^{\mathrm{univ}}$ | the projective bundle to $\mathbf{E}^{\mathrm{univ}}$, 68 |
| $\mathfrak{E}^{\mathrm{univ}}_{\mathfrak{a}}$ | the completion of $\bar{\mathbf{E}}^{\mathrm{univ}}_{\mathfrak{a}}$, 68 |
| $\mathfrak{N}^{d-1}_I$ | part of the formal boundary, 69 |
| $\mathfrak{M}^{d-1,1}_I$, $M^{d-1,1}_I$, $\bar{M}^{d-1,1}_I$ | induced schemes, 71 |
| $u_\Lambda$ | the exponential series to a lattice, 73 |
| $\mathfrak{E}/\mathfrak{a}$ | completion of $\overline{\mathbf{E}/\mathfrak{a}}$ 75 |
| $\mathbf{T}$ | the universal Drinfeld module with bad reduction, 78 |
| $\mathfrak{P}^1_A$ | the completion of $\mathbb{P}^1_A$, 79 |
| $G\times^H X$ | the induced scheme, 85 |

# Index


additive module, 17
additive polynomial, 4

Betti cohomology, 38

canonical factorization, 60
categorical quotient, 8
characteristic
  away from, 20
  generic, 23
  of a Drinfeld module, 20
completion, 51, 52
congruence subgroup, 49
convergence lemma, 10

de Rham cohomology, 39
defect, 41
deformation, 33, 42
  isomorphism of, 33
degree
  of an isogeny, 21
degree of a homomorphism, 6
discrete set, 15, 57
division points, 25
divisor
  integral part, 61
Drinfeld module, 16, 17, 19
  characteristic of, 20
  good reduction, 54
  having coefficients in $O$, 54
  homomorphism of, 21
  projective bundle, 68
  rank of, 21
  standard, 21
  with level $I$ structure, 27

endomorphism, 3

finite type, 52
formal boundary, 53
formal scheme, 51
  affine, 51
  of finite type, 52
Frobenius, 2

geometric quotient, 7
good reduction, 54

height, 22
homomorphism, 3, 21, 22
  of Drinfeld modules, 17

induced scheme, 85
isogeny, 21

kernel, 22

lattice, 15, 57
  universal, 73
level structure, 27
  lattice, 57
  normal, 62
  total, 27
line bundle, 1, 3

morphism
  of a lattice, 15
  of Drinfeld modules, 27
  of Tate data, 58

principal homogeneous space, 8

quotient, 7

rank, 58
  of a Drinfeld module, 21

section at infinity, 68

skew-polynomial ring, 4
small extension, 34
stable reduction, 54
standard, 21
standard homomorphism, 6

tangent space, 35
Tate algebra, 14
Tate datum, 66
  normal, 58
Tate uniformization, 60

## Editorial Information

To be published in the *Memoirs*, a paper must be correct, new, nontrivial, and significant. Further, it must be well written and of interest to a substantial number of mathematicians. Piecemeal results, such as an inconclusive step toward an unproved major theorem or a minor variation on a known result, are in general not acceptable for publication.

Papers appearing in *Memoirs* are generally at least 80 and not more than 200 published pages in length. Papers less than 80 or more than 200 published pages require the approval of the Managing Editor of the Transactions/Memoirs Editorial Board.

As of September 30, 2008, the backlog for this journal was approximately 15 volumes. This estimate is the result of dividing the number of manuscripts for this journal in the Providence office that have not yet gone to the printer on the above date by the average number of monographs per volume over the previous twelve months, reduced by the number of volumes published in four months (the time necessary for preparing a volume for the printer). (There are 6 volumes per year, each usually containing at least 4 numbers.)

A Consent to Publish and Copyright Agreement is required before a paper will be published in the *Memoirs*. After a paper is accepted for publication, the Providence office will send a Consent to Publish and Copyright Agreement to all authors of the paper. By submitting a paper to the *Memoirs*, authors certify that the results have not been submitted to nor are they under consideration for publication by another journal, conference proceedings, or similar publication.

### Information for Authors

*Memoirs* are printed from camera copy fully prepared by the author. This means that the finished book will look exactly like the copy submitted.

**Initial submission.** The AMS uses Centralized Manuscript Processing for initial submissions. Authors should submit a PDF file using the Initial Manuscript Submission form found at `www.ams.org/peer-review-submission`, or send one copy of the manuscript to the following address: Centralized Manuscript Processing, MEMOIRS OF THE AMS, 201 Charles Street, Providence, RI 02904-2294 USA. If a paper copy is being forwarded to the AMS, indicate that it is for it Memoirs and include the name of the corresponding author, contact information such as email address or mailing address, and the name of an appropriate Editor to review the paper (see the list of Editors below).

The paper must contain a *descriptive title* and an *abstract* that summarizes the article in language suitable for workers in the general field (algebra, analysis, etc.). The *descriptive title* should be short, but informative; useless or vague phrases such as "some remarks about" or "concerning" should be avoided. The *abstract* should be at least one complete sentence, and at most 300 words. Included with the footnotes to the paper should be the 2000 *Mathematics Subject Classification* representing the primary and secondary subjects of the article. The classifications are accessible from `www.ams.org/msc/`. The list of classifications is also available in print starting with the 1999 annual index of *Mathematical Reviews*. The Mathematics Subject Classification footnote may be followed by a list of *key words and phrases* describing the subject matter of the article and taken from it. Journal abbreviations used in bibliographies are listed in the latest *Mathematical Reviews* annual index. The series abbreviations are also accessible from `www.ams.org/msnhtml/serials.pdf`. To help in preparing and verifying references, the AMS offers MR Lookup, a Reference Tool for Linking, at `www.ams.org/mrlookup/`.

**Electronically prepared manuscripts.** The AMS encourages electronically prepared manuscripts, with a strong preference for $\mathcal{AMS}$-LaTeX. To this end, the Society has prepared $\mathcal{AMS}$-LaTeX author packages for each AMS publication. Author packages include instructions for preparing electronic manuscripts, samples, and a style file that generates the particular design specifications of that publication series. Though $\mathcal{A}\mathcal{M}\mathcal{S}$-LaTeX is the highly preferred format of TeX, author packages are also available in $\mathcal{A}\mathcal{M}\mathcal{S}$-TeX.

Authors may retrieve an author package for *Memoirs of the AMS* from `www.ams.org/journals/memo/memoauthorpac.html` or via FTP to `ftp.ams.org` (login as `anonymous`, enter username as password, and type `cd pub/author-info`). The *AMS Author Handbook* and the *Instruction Manual* are available in PDF format from the author package link. The author package can also be obtained free of charge by sending email to `tech-support@ams.org` (Internet) or from the Publication Division, American Mathematical Society, 201 Charles St., Providence, RI 02904-2294, USA. When requesting an author package, please specify $\mathcal{A}\mathcal{M}\mathcal{S}$-LaTeX or $\mathcal{A}\mathcal{M}\mathcal{S}$-TeX and the publication in which your paper will appear. Please be sure to include your complete mailing address.

**After acceptance.** The final version of the electronic file should be sent to the Providence office (this includes any TeX source file, any graphics files, and the DVI or PostScript file) immediately after the paper has been accepted for publication.

Before sending the source file, be sure you have proofread your paper carefully. The files you send must be the EXACT files used to generate the proof copy that was accepted for publication. For all publications, authors are required to send a printed copy of their paper, which exactly matches the copy approved for publication, along with any graphics that will appear in the paper.

Accepted electronically prepared files can be submitted via the web at `www.ams.org/submit-book-journal/`, sent via FTP, or sent on CD-Rom or diskette to the Electronic Prepress Department, American Mathematical Society, 201 Charles Street, Providence, RI 02904-2294 USA. TeX source files, DVI files, and PostScript files can be transferred over the Internet by FTP to the Internet node `ftp.ams.org` (130.44.1.100). When sending a manuscript electronically via CD-Rom or diskette, please be sure to include a message identifying the paper as a Memoir.

Electronically prepared manuscripts can also be sent via email to `pub-submit@ams.org` (Internet). In order to send files via email, they must be encoded properly. (DVI files are binary and PostScript files tend to be very large.)

**Electronic graphics.** Comprehensive instructions on preparing graphics are available at `www.ams.org/authors/journals.html`. A few of the major requirements are given here.

Submit files for graphics as EPS (Encapsulated PostScript) files. This includes graphics originated via a graphics application as well as scanned photographs or other computer-generated images. If this is not possible, TIFF files are acceptable as long as they can be opened in Adobe Photoshop or Illustrator. No matter what method was used to produce the graphic, it is necessary to provide a paper copy to the AMS.

Authors using graphics packages for the creation of electronic art should also avoid the use of any lines thinner than 0.5 points in width. Many graphics packages allow the user to specify a “hairline” for a very thin line. Hairlines often look acceptable when proofed on a typical laser printer. However, when produced on a high-resolution laser imagesetter, hairlines become nearly invisible and will be lost entirely in the final printing process.

Screens should be set to values between 15% and 85%. Screens which fall outside of this range are too light or too dark to print correctly. Variations of screens within a graphic should be no less than 10%.

**Inquiries.** Any inquiries concerning a paper that has been accepted for publication should be sent to `memo-query@ams.org` or directly to the Electronic Prepress Department, American Mathematical Society, 201 Charles St., Providence, RI 02904-2294 USA.

## Editors

This journal is designed particularly for long research papers, normally at least 80 pages in length, and groups of cognate papers in pure and applied mathematics. Papers intended for publication in the *Memoirs* should be addressed to one of the following editors. The AMS uses Centralized Manuscript Processing for initial submissions to AMS journals. Authors should follow instructions listed on the Initial Submission page found at www.ams.org/memo/memosubmit.html.

**Algebra** to ALEXANDER KLESHCHEV, Department of Mathematics, University of Oregon, Eugene, OR 97403-1222; email: ams@noether.uoregon.edu

**Algebraic geometry and its application** to MINA TEICHER, Emmy Noether Research Institute for Mathematics, Bar-Ilan University, Ramat-Gan 52900, Israel; email: teicher@macs.biu.ac.il

**Algebraic geometry** to DAN ABRAMOVICH, Department of Mathematics, Brown University, Box 1917, Providence, RI 02912; email: amsedit@math.brown.edu

**Algebraic topology** to ALEJANDRO ADEM, Department of Mathematics, University of British Columbia, Room 121, 1984 Mathematics Road, Vancouver, British Columbia, Canada V6T 1Z2; email: adem@math.ubc.ca

**Combinatorics** to JOHN R. STEMBRIDGE, Department of Mathematics, University of Michigan, Ann Arbor, Michigan 48109-1109; email: FRS@umich.edu

**Complex analysis and harmonic analysis** to ALEXANDER NAGEL, Department of Mathematics, University of Wisconsin, 480 Lincoln Drive, Madison, WI 53706-1313; email: nagel@math.wisc.edu

**Differential geometry and global analysis** to LISA C. JEFFREY, Department of Mathematics, University of Toronto, 100 St. George St., Toronto, ON Canada M5S 3G3; email: jeffrey@math.toronto.edu

**Dynamical systems and ergodic theory and complex anaysis** to YUNPING JIANG, Department of Mathematics, CUNY Queens College and Graduate Center, 65-30 Kissena Blvd., Flushing, NY 11367; email: Yunping.Jiang@qc.cuny.edu

**Functional analysis and operator algebras** to DIMITRI SHLYAKHTENKO, Department of Mathematics, University of California, Los Angeles, CA 90095; email: shlyakht@math.ucla.edu

**Geometric analysis** to WILLIAM P. MINICOZZI II, Department of Mathematics, Johns Hopkins University, 3400 N. Charles St., Baltimore, MD 21218; email: trans@math.jhu.edu

**Geometric analysis** to MARK FEIGHN, Math Department, Rutgers University, Newark, NJ 07102; email: feighn@andromeda.rutgers.edu

**Harmonic analysis, representation theory, and Lie theory** to ROBERT J. STANTON, Department of Mathematics, The Ohio State University, 231 West 18th Avenue, Columbus, OH 43210-1174; email: stanton@math.ohio-state.edu

**Logic** to STEFFEN LEMPP, Department of Mathematics, University of Wisconsin, 480 Lincoln Drive, Madison, Wisconsin 53706-1388; email: lempp@math.wisc.edu

**Number theory** to JONATHAN ROGAWSKI, Department of Mathematics, University of California, Los Angeles, CA 90095; email: jonr@math.ucla.edu

**Partial differential equations** to GUSTAVO PONCE, Department of Mathematics, South Hall Room 6607, University of California, Santa Barbara, CA 93106; email: ponce@math.ucsb.edu

**Partial differential equations and dynamical systems** to PETER POLACIK, School of Mathematics, University of Minnesota, Minneapolis, MN 55455; email: polacik@math.umn.edu

**Probability and statistics** to RICHARD BASS, Department of Mathematics, University of Connecticut, Storrs, CT 06269-3009; email: bass@math.uconn.edu

**Real analysis and partial differential equations** to DANIEL TATARU, Department of Mathematics, University of California, Berkeley, Berkeley, CA 94720; email: tataru@math.berkeley.edu

**All other communications to the editors** should be addressed to the Managing Editor, ROBERT GURALNICK, Department of Mathematics, University of Southern California, Los Angeles, CA 90089-1113; email: guralnic@math.usc.edu.

## Titles in This Series


923 **Michael Jöllenbeck and Volkmar Welker,** Minimal resolutions via algebraic discrete Morse theory, 2009

922 **Ph. Barbe and W. P. McCormick,** Asymptotic expansions for infinite weighted convolutions of heavy tail distributions and applications, 2009

921 **Thomas Lehmkuhl,** Compactification of the Drinfeld modular surfaces, 2009

920 **Georgia Benkart, Thomas Gregory, and Alexander Premet,** The recognition theorem for graded Lie algebras in prime characteristic, 2009

919 **Roelof W. Bruggeman and Roberto J. Miatello,** Sum formula for $SL_2$ over a totally real number field, 2009

918 **Jonathan Brundan and Alexander Kleshchev,** Representations of shifted Yangians and finite $W$-algebras, 2008

917 **Salah-Eldin A. Mohammed, Tusheng Zhang, and Huaizhong Zhao,** The stable manifold theorem for semilinear stochastic evolution equations and stochastic partial differential equations, 2008

916 **Yoshikata Kida,** The mapping class group from the viewpoint of measure equivalence theory, 2008

915 **Sergiu Aizicovici, Nikolaos S. Papageorgiou, and Vasile Staicu,** Degree theory for operators of monotone type and nonlinear elliptic equations with inequality constraints, 2008

914 **E. Shargorodsky and J. F. Toland,** Bernoulli free-boundary problems, 2008

913 **Ethan Akin, Joseph Auslander, and Eli Glasner,** The topological dynamics of Ellis actions, 2008

912 **Igor Chueshov and Irena Lasiecka,** Long-time behavior of second order evolution equations with nonlinear damping, 2008

911 **John Locker,** Eigenvalues and completeness for regular and simply irregular two-point differential operators, 2008

910 **Joel Friedman,** A proof of Alon's second eigenvalue conjecture and related problems, 2008

909 **Cameron McA. Gordon and Ying-Qing Wu,** Toroidal Dehn fillings on hyperbolic 3-manifolds, 2008

908 **J.-L. Waldspurger,** L'endoscopie tordue n'est pas si tordue, 2008

907 **Yuanhua Wang and Fei Xu,** Spinor genera in characteristic 2, 2008

906 **Raphaël S. Ponge,** Heisenberg calculus and spectral theory of hypoelliptic operators on Heisenberg manifolds, 2008

905 **Dominic Verity,** Complicial sets characterising the simplicial nerves of strict $\omega$-categories, 2008

904 **William M. Goldman and Eugene Z. Xia,** Rank one Higgs bundles and representations of fundamental groups of Riemann surfaces, 2008

903 **Gail Letzter,** Invariant differential operators for quantum symmetric spaces, 2008

902 **Bertrand Toën and Gabriele Vezzosi,** Homotopical algebraic geometry II: Geometric stacks and applications, 2008

901 **Ron Donagi and Tony Pantev (with an appendix by Dmitry Arinkin),** Torus fibrations, gerbes, and duality, 2008

900 **Wolfgang Bertram,** Differential geometry, Lie groups and symmetric spaces over general base fields and rings, 2008

899 **Piotr Hajłasz, Tadeusz Iwaniec, Jan Malý, and Jani Onninen,** Weakly differentiable mappings between manifolds, 2008

898 **John Rognes,** Galois extensions of structured ring spectra/Stably dualizable groups, 2008

897 **Michael I. Ganzburg,** Limit theorems of polynomial approximation with exponential weights, 2008

896 **Michael Kapovich, Bernhard Leeb, and John J. Millson,** The generalized triangle inequalities in symmetric spaces and buildings with applications to algebra, 2008

895 **Steffen Roch,** Finite sections of band-dominated operators, 2008

894 **Martin Dindoš,** Hardy spaces and potential theory on $C^1$ domains in Riemannian manifolds, 2008

893 **Tadeusz Iwaniec and Gaven Martin,** The Beltrami Equation, 2008

892 **Jim Agler, John Harland, and Benjamin J. Raphael,** Classical function theory, operator dilation theory, and machine computation on multiply-connected domains, 2008

891 **John H. Hubbard and Peter Papadopol,** Newton's method applied to two quadratic equations in $\mathbb{C}^2$ viewed as a global dynamical system, 2008

890 **Steven Dale Cutkosky,** Toroidalization of dominant morphisms of 3-folds, 2007

889 **Michael Sever,** Distribution solutions of nonlinear systems of conservation laws, 2007

888 **Roger Chalkley,** Basic global relative invariants for nonlinear differential equations, 2007

887 **Charlotte Wahl,** Noncommutative Maslov index and eta-forms, 2007

886 **Robert M. Guralnick and John Shareshian,** Symmetric and alternating groups as monodromy groups of Riemann surfaces I: Generic covers and covers with many branch points, 2007

885 **Jae Choon Cha,** The structure of the rational concordance group of knots, 2007

884 **Dan Haran, Moshe Jarden, and Florian Pop,** Projective group structures as absolute Galois structures with block approximation, 2007

883 **Apostolos Beligiannis and Idun Reiten,** Homological and homotopical aspects of torsion theories, 2007

882 **Lars Inge Hedberg and Yuri Netrusov,** An axiomatic approach to function spaces, spec tral synthesis and Luzin approximation, 2007

881 **Tao Mei,** Operator valued Hardy spaces, 2007

880 **Bruce C. Berndt, Geumlan Choi, Youn-Seo Choi, Heekyoung Hahn, Boon Pin Yeap, Ae Ja Yee, Hamza Yesilyurt, and Jinhee Yi,** Ramanujan's forty identities for Rogers-Ramanujan functions, 2007

879 **O. García-Prada, P. B. Gothen, and V. Muñoz,** Betti numbers of the moduli space of rank 3 parabolic Higgs bundles, 2007

878 **Alessandra Celletti and Luigi Chierchia,** KAM stability and celestial mechanics, 2007

877 **María J. Carro, José A. Raposo, and Javier Soria,** Recent developments in the theory of Lorentz spaces and weighted inequalities, 2007

876 **Gabriel Debs and Jean Saint Raymond,** Borel liftings of Borel sets: Some decidable and undecidable statements, 2007

875 **C. Krattenthaler and T. Rivoal,** Hypergéométrie et fonction zêta de Riemann, 2007

874 **Sonia Natale,** Semisolvability of semisimple Hopf algebras of low dimension, 2007

873 **A. J. Duncan,** Exponential genus problems in one-relator products of groups, 2007

872 **Anthony V. Geramita, Tadahito Harima, Juan C. Migliore, and Yong Su Shin,** The Hilbert function of a level algebra, 2007

871 **Pascal Auscher,** On necessary and sufficient conditions for $L^p$-estimates of Riesz transforms associated to elliptic operators on $\mathbb{R}^n$ and related estimates, 2007

870 **Takuro Mochizuki,** Asymptotic behaviour of tame harmonic bundles and an application to pure twistor $D$-modules, Part 2, 2007

869 **Takuro Mochizuki,** Asymptotic behaviour of tame harmonic bundles and an application to pure twistor $D$-modules, Part 1, 2007